Blue Light Responses:

Phenomena and Occurrence in Plants and Microorganisms

Volume I

Editor

Horst Senger, Dr. rer.nat.
Professor of Botany
Department of Biology/Botany
Philipps-University
Marburg
West Germany

CRC Press, Inc.
Boca Raton, Florida

Library of Congress Cataloging-in-Publication Data

Blue light responses.
  Includes bibliographies and indexes.
  1. Blue light—Physiological effect. 2. Plants,
Effect of blue light on. 3. Ultraviolet radiation —
Physiological effect. I. Senger, H. (Horst), 1931-
QH515.B569 1987   574.19′154   86-8325
ISBN 0-8493-5235-5 (v. 1)
ISBN 0-8493-5236-3 (v. 2)

Direct all inquiries to CRC Press, Inc., 2000 Corporate Blvd., N.W., Boca Raton, Florida, 33431.

© 1987 by CRC Press, Inc.

International Standard Book Number 0-8493-5235-5 (Volume I)
International Standard Book Number 0-8493-5236-3 (Volume II)

Library of Congress Card Number 86-8325
Printed in the United States

THE EDITOR

Horst Senger, Dr. rer.nat., is Professor of Botany, Department of Botany, Philipps-University Marburg.

He received his doctoral degree from the University of Göttingen and his Habilitation from the University of Marburg. He has held postdoctoral positions at the University of Tübingen and at Oregon State University and has been Guest Scientist at the University of Natal, University of Tokyo, Oregon State University, and at the CSIRO in Canberra.

Professor Senger has worked on synchronization of microalgae, chloroplast development, chlorophyll biosynthesis, photohydrogen evolution, and blue light effects on algae. His work has been published in over 120 original papers, and he is co-editor of a book *Regulation of Chloroplast Differentiation*. He has organized two International Conferences on the ''Effect of Blue Light on Plants and Microorganisms'' and edited two books, *The Blue Light Syndrome* and *Blue Light Effects in Biological Systems*.

CONTRIBUTORS

Volume I

Helga Drumm-Herrel, Dr. rer.nat.
Akademischer Oberrat
Biological Institute II
University of Freiburg
Freiburg
West Germany

Roger Durand, Dr.
Fungal Differentiation Laboratory
University of Lyon I
Villeurbanne
France

Donat P. Häder, Dr.
Dozent
Department of Biology
Philipps University
Marburg
West Germany

Wolfgang Kowallik, Dr. rer.nat.
Professor
Faculty of Biology
University of Bielefeld
Bielefeld
West Germany

Hans Mohr, Dr. rer.nat.
Professor
Biological Institute II
University of Freiburg
Freiburg
West Germany

Helga Ninnemann, Dr. phil.nat.
Professor
Faculty of Chemistry
University of Tübingen
Tübingen
West Germany

W. Rau, Dr.
Professor
Botanical Institute
University of Munich
Munich
West Germany

Rainer Schmid, Dr. rer.nat.
Dozent
Institute of Plant Physiology
  Cell Biology, and Microbiology
Free University
Berlin
West Germany

Erich L. Schrott, Dr. rer.nat.
Dozent
Botanical Institute
University of Munich
Munich
West Germany

Horst Senger, Dr. rer.nat.
Professor
Department of Biology/Botany
Phillips-University
Marburg, West Germany

CONTRIBUTORS

Volume II

Matthew J. Dring, Ph.D.
Department of Botany
The Queen's University of Belfast
Belfast
Northern Ireland

Paul A. Galland, Dr.
Research Assistant Professor
Department of Physics
Syracuse University
Syracuse, New York

Jonathan B. Gressel, Ph.D.
Professor
Department of Plant Genetics
The Weizman Institute of Science
Rehovot
Israel

Benjamin A. Horwitz, Ph.D.
Postdoctoral Fellow
Department of Plant Biology
Carnegie Institution of Washington
Stanford, California

Christer Larsson, Ph.D.
Assistant Professor
Department of Plant Physiology
University of Lund
Lund
Sweden

Günter Ruyters, Ph.D.
Assistant Professor
Faculty of Biology IV
University of Bielefeld
Bielefeld
West Germany

Werner Schmidt, Dr.
Docent
Faculty of Biology
University of Konstanz
Konstanz
West Germany

Horst Senger, Dr.
Professor
Department of Botany
Philipps-University of Marburg
Marburg
West Germany

Pill-Soon Song, Ph.D.
Professor
Department of Chemistry
University of Nebraska
Lincoln, Nebraska

Susanne Widell, Ph.D.
Department of Plant Physiology
University of Lund
Lund
Sweden

Eduardo Zeiger, Ph.D.
Professor
Department of Biological Sciences
Stanford University
Stanford, California

TABLE OF CONTENTS

Volume I

Introduction

Chapter 1

INTRODUCTION

Horst Senger

All life on earth depends directly or indirectly on light. We have to distinguish principally two modes of action by which light is used: as a source of energy and as a signal for photomorphogenetic processes in the broadest sense. By far the most important process for converting light energy into chemical energy is photosynthesis. The light requirement of this process is characterized by very high energy. To the best of our knowledge the pigments involved in photosynthesis do not act as photoreceptors for photomorphogenetic effects. However, there are some cases in which photosynthetic products as substrates indirectly regulate developmental processes. But these cases should not be considered photomorphogenetic events.

Pure photomorphogenetic effects are generally characterized by low energy requirements of radiation. There are processes which need only a short light pulse to trigger the chain of subsequent reactions, and other processes which need a permanent input of light in order to maintain a certain state of reactivity or to provide a particular potential for a reaction.

Parts of almost the whole visible spectrum can be used by one or another organism as signals for photomorphogenetic responses. Thus the number of photoreceptors involved is accordingly high.

Among the various photoreceptors those for absorbing blue light are probably the oldest phylogenetically, but, nevertheless, they are the ones least understood. No specific blue light photoreceptor has been unequivocally identified or isolated, although flavoproteins and carotenoproteins are the most favored contenders.[1-3] The reactions mediated by blue light can be classified as photomorphogenetic effects and photomovement. Under this aspect, photomorphogenetic effects include morphological, metabolic, and adaptational processes.

In some cases it is difficult to draw the boundary between actinic light for pure blue light effects and the green ultraviolet (UV) part of the spectrum. In particular, the border towards the near UV (UV-A, 320 to 400 nm) fluctuates. Therefore, the responsible light receptors are often referred to as blue/UV photoreceptors. A common action spectrum for blue light effects exhibits its main peak at 450 nm and peaks or shoulders around 480, 420, and in addition, sometimes at 370 nm. The photoreceptor responsible for such an action spectrum is often referred to as "cryptochrome". However, many published action spectra of blue light effects deviate from this type. Even taking into account the possibilities of the distortion of action spectra by screening and "antenna pigments", one has to expect a variety of different blue-light photoreceptors.

Observations, interpretable as "blue light responses", were already reported in the middle of the last century,[4] but the first specific investigation of blue light effects dates back to the early 1930s, when phototropic curvature was studied (see Galston[5]) and the first precise action spectrum of this effect was published.[6]

For a long time, interest in the action of blue light was focused on phototropism and related processes like photooxidation of plant growth hormones.

The coincidence of technical advances in the production of high-energy light sources, and improvements in monochromators and light detectors coupled with the discovery that blue light regulates carbon metabolism[7-9] opened a new era of blue light research. Several review articles have covered the advancements. Up to 1981, they are quoted by Senger and Briggs.[10] Subsequent literature is covered by the reviews of Kowallik,[11] Senger,[12] Briggs and Iino,[13] Gressel and Rau,[14] Dörnemann and Senger,[15] and Schmidt.[16] Two international meetings (in 1979 and 1983) have been entirely devoted to blue light effects in plants and microorganisms.[17,18]

The first part of the present book is devoted to the documentation of the broad variety of blue light effects. Phototropism as the classical subject of blue light physiology has recently been covered by Dennison[19] and is therefore omitted here. The other extensively studied topics of blue light research are presented in an updated manner. They are supplemented by new findings studied in more recently introduced organisms. With respect to their in vivo significance, the evaluation of various blue light effects, often studied in isolated systems, requires the consideration of interrelated light effects. Among those, phytochrome-controlled processes and photosynthesis as the most important light-mediated reactions are covered.

A question as old as blue light research itself is one regarding the photoreceptor pigments and the mode of signal transduction. A good part of the book is devoted to these issues, including some critical considerations of methodological problems. The last chapter deals with the more recently evolving interest in the ecological relevance of blue light effects.

REFERENCES

1. **Song, P. S.**, Spectroscopic and photochemical characterization of flavoproteins and carotenoproteins as blue light photoreceptors, in *The Blue Light Syndrome*, Senger, H., Ed., Springer-Verlag, Berlin, 1980, 157.
2. **Shropshire, W., Jr.**, Carotenoids as primary photoreceptors, in *The Blue Light Syndrome*, Senger, H., Ed., Springer-Verlag, Berlin, 1980, 172.
3. **De Fabo, E.**, On the nature of the blue light photoreceptor: still an open question, *The Blue Light Syndrome*, Senger, H., Ed., Springer-Verlag, Berlin, 1980, 187.
4. **Sachs, V. J.**, Wirkungen des farbigen Lichts auf Pflanzen, *Bot. Z.*, 22, 353, 1864.
5. **Galston, A. W.**, Phototropism of stems, roots and coleoptiles, *Encycl. Plant Physiol.*, 17, 492, 1959.
6. **Johnston, E. S.**, Phototropic sensitivity in relation to wavelength, *Smithson. Misc. Collect.*, 92, 1, 1934.
7. **Hauschild, A. H. W., Nelson, C. D., and Krokov, G.**, Concurrent changes in the products and the rate of photosynthesis in *Chlorella vulgaris* in the presence of blue light, *Naturwissenschaften*, 51, 274, 1964.
8. **Kowallik, W.**, Über die Wirkung des blauen und roten Spektralbereichs auf die Zusammensetzung und Zellteilung synchronisierter Chlorellen, *Planta*, 58, 337, 1962.
9. **Voskresenskaya, N. P.**, The importance of spectral composition of the light for photosynthetic formation of substances, *Dokl. Acad. Nauk SSSR*, 93, 911, 1953.
10. **Senger, H. and Briggs, W.**, The blue light receptor(s): primary reactions and subsequent metabolic changes, in *Photochemical and Photobiological Reviews*, Vol. 8, Smith, K. C., Ed., Plenum Press, New York, 1981, 1.
11. **Kowallik, W.**, Blue light effects on respiration, *Annu. Rev. Plant Physiol.*, 33, 51, 1982.
12. **Senger, H.**, The effect of blue light on plants and microorganisms, *Photochem. Photobiol.*, 35, 911, 1982.
13. **Briggs, W. R. and Iino, M.**, Blue-light-absorbing photoreceptors in plants, *Phil. Trans. R. Soc. London*, B303, 345, 1983.
14. **Gressel, J. and Rau, W.**, Photocontrol of fungal development, photomorphogenesis, in *Encyclopedia of Plant Physiology, New Series*, Vol. 16, Shropshire, W., Jr. and Mohr, H., Eds., Springer-Verlag, Berlin, 1983, 603.
15. **Dörnemann, D. and Senger, H.**, *Blue Light Photoreceptors, Techniques in Photomorphogenesis*, Smith, H. and Holmes, G., Eds., 1984, 279.
16. **Schmidt, W.**, Bluelight physiology, *BioScience*, 34, 698, 1984.
17. **Senger, H., Ed.**, *The Blue Light Syndrome*, Springer-Verlag, Berlin, 1980.
18. **Senger, H., Ed.**, *Blue Light Effects in Biological Systems*, Springer-Verlag, Berlin, 1984.
19. **Dennison, D. S.**, phototropism, physiology of movements, in *Encyclopedia of Plant Physiology*, New Series, Vol. 7, Haupt, W. and Feinlieb, M. E., Eds., Springer-Verlag, Berlin, 1979, 506.

Phenomena, Distribution, and Mechanisms of Blue Light Effects

---

Chapter 2

BLUE LIGHT EFFECTS ON CARBOHYDRATE AND PROTEIN METABOLISM

Wolfgang Kowallik

TABLE OF CONTENTS

I. INTRODUCTION

The visible electromagnetic radiation — light — contains photons of various wavelengths and energy content ranging from about 350 to 750 nm or 5.8 to 2.7×10^{-12} ergs, respectively, which cover the light colors of the rainbow from ultraviolet (UV) to dark red. Living matter, on the other hand, contains substances which selectively absorb parts of that radiation; e.g., they interfere with specific light quanta. Therefore, it is easy to see why for a long time scientists have investigated whether there are specific influences of various parts of the spectrum on living organisms. The discovery of the wavelength dependences of photosynthesis, of phototropism, of photoperiodism, and of photomorphogenesis are important results of such efforts. There also have been attempts to find out whether there are influences of specific wavelengths of light on the basic composition of the cell.

II. INFLUENCE OF LIGHT QUALITY ON THE LEVEL OF CARBOHYDRATE AND OF PROTEIN IN VARIOUS ORGANISMS

Differences in the content of carbohydrates and of protein in green organisms exposed to various spectral regions have been reported in the last century. However, these results are without significance, since the light applied did not only vary in color but also in intensity. The first reports which took this into account came from a Russian group some 30 years ago. In 1952 to 1953 Voskresenskaya[1,2] reported greater amounts of nitrogenous substances in leaves and/or seeds of tobacco, sunflower, corn, and bean when these plants were exposed to blue instead of red light. Growth in red light resulted in a greater accumulation of carbohydrates. In the following years, these results were proved not only for other higher plants, but also for ferns and various algae. Appleman and Pyfrom[3] showed it for barley and some other cultivated plants, Howell et al.[4] for soybean, and Casper and Pirson[5] for the succulent CAM-plant *Kalanchoe*. Gametophytes of *Dryopteris filix-mas* responded likewise,[6-10] as did those of *Pteridium aquilinum*.[11] Several algae have been examined, especially in Pirson's laboratory, in which most work has been done with *Chlorella*. This alga was synchronized by a regime of 14 hr blue/12 hr dark or 14 hr red/12 hr dark, adjusted at both light qualities to yield an equal dry-weight production. There was complete cell division in each dark period, after which the cultures were diluted to a density of 1.56×10^6 cells per milliliter. By this procedure the algae could be kept growing continuously in the respective light color for several months. At all developmental stages of the cell a difference in the basic composition was found. It was most pronounced, however, in the middle of the light period. At that time the dry matter of cells which were exposed to blue light contained only 15% carbohydrate but 60% protein. Those grown in red light contained more than twice this amount of carbohydrate (39%), but with only 29% protein.[12,13] The wavelength dependence for the increase in the amount of protein revealed maxima around 460 and 370 nm.[14] These results were basically confirmed several times.[15-17] Beyond *Chlorella*, the same effect of the color of the light used for autotrophic growth was observed with *Euglena*,[15] with *Chlorogonium*,[18] with *Chlamydomonas*,[19] and with the diatom *Cyclotella*.[15] At the same time Clauss[20,21] discovered it in the siphonaceous green alga *Acetabularia;* later it was seen in *Scenedesmus* by Brinkmann and Senger,[22] and most recently in *Spirulina* by Jeeji Bai and Subramanian.[23]

III. IS THERE AN EFFECT OF BLUE LIGHT ON THE PRODUCTS OF PHOTOSYNTHESIS?

The above difference in the wavelength-dependent basic composition of the plant cell would result, if there were different products of photosynthesis in both light qualities. The fate of photosynthetically incorporated carbon has already been examined by Voskresenskaya[1] and later by Cayle and Emerson[24] with the green alga *Chlorella*. Using radioactive $NaH^{14}CO_3$ the latter authors found an almost equal total incorporation of carbon at equal fluence rates of light of $\lambda = 436$ nm and of $\lambda = 644$ nm, but a different distribution of label in the fixation products. After 30 sec of blue light there was more radioactivity detectable in the amino acids glycine, serine, and alanine than after the same time of red light exposure. In blue light the respective specific activities were 2.5, 1.9, and 3.0 times greater than in red light. Furthermore, there was a different distribution of label within the glycine molecule. While in red light about 80% of the total radioactivity was found in the carboxyl- and only 20% in the α-carbon atom, in blue light this distribution was 60:40. This has been discussed as being indicative of different pathways for glycine production, perhaps glycolate being the precursor in blue light.

A positive effect of blue light on the formation of serine and alanine, but in addition of aspartic, glutamic, malic, and fumaric acids, was also found by Hauschild et al.,[25,26] when they allowed *Chlorella* cells to photosynthesize in broad fields of blue or of red light for longer periods of time. Their samples were usually taken after 30 min of photosynthesis. Essentially the same results were obtained when red light was supplemented by as little as 4% of blue light. Therefore, the latter spectral region appears to create some specific conditioning of the metabolism of the cell,[27,28] rather than to drive a different primary photosynthetic reaction. The increased production of malic, aspartic, and glutamic acid might derive from an influence on the carboxylation of phosphoenolpyruvate. Indeed, a greater nonphotosynthetic C fixation in that latter substance has been reported at 3-(4-chlorophenyl)-1-1-dimethylurea (CMU) poisoning by Ogasawara and Miyachi[29,30] and by using chlorophyll-free *Chlorella* mutants by Kamiya and Miyachi[31] and by Ruyters[32] on exposure and even after preexposure to blue light. So there is not much evidence for a direct effect of blue light on the primary products of photosynthesis, but rather there appears to exist a blue-light-dependent regulatory system, perhaps draining off C_3 compounds from the Calvin cycle but also from glycolysis.

IV. ENHANCEMENT OF CARBOHYDRATE DEGRADATION BY BLUE LIGHT

A greater percentage of protein of the dry matter of an organism would also result if the biosynthesis of protein were enhanced leading to a decrease in carbohydrate by draining off degradation products of the latter. It would, however, also occur, if blue light would enhance carbohydrate degradation, thus delivering more intermediates for the production of amino acids and protein. Since on blue light exposure a green cell will photosynthesize, this metabolism will interfere with that of respiration. Because of that, the light effect on carbon metabolism has first been examined under conditions without concurrent photosynthesis, and only thereafter was an attempt made to prove it at concurrent photosynthesis.

A. Without Concurrent Photosynthesis
Photosynthesis can be excluded by specific poisoning or by a suitable mutation. Both methods have been applied. When autotrophically grown *Chlorella* cells were inhibited by the addition of 10^{-5} mol DCMU 1^{-1} to their normal growth medium, they lost 67.5% of their reserve carbohydrates in blue light, but only 54.8% in darkness within 5 hr. The same ratio was also found when the above cells were suspended in plain phosphate buffer (57.2:44.2%). This latter result points to carbohydrate degradation as the target of blue light action and largely rejects the above idea of a blue light effect on protein biosynthesis. Consequently the respiratory gas exchange has been followed. Oxygen consumption and carbon dioxide liberation were both found to be enhanced in blue light, leaving the RQ unchanged at about unity.[33] For a closer characterization of the response, the intensity and

wavelength dependences for the enhancement of O_2 uptake have been determined. The effect was already saturated at the low fluence rate of about 300 $\mu W \cdot cm^{-2}$. The wavelength dependence revealed maxima around 380 and 460 nm and no effect, whatsoever, of green, yellow, red, and far-red light.[34] This spectrum matches that of the enhanced protein production mentioned above[14] and the comparable one at DCMU poisoning,[35] indicating a close correlation of both processes. From the course of these spectra the effect of light is mediated by a pigment absorbing like flavins or *cis*-carotenoids. That there is no specific effect of blue-excited chlorophyll, indeed, has been shown by the use of pigment mutants of *Chlorella*. Being chlorophyll free, one of them is bright yellow because of abandoned carotenoids, and the other one is white, since it additionally lacks the yellow pigments. Both organisms have to be grown with glucose as an exogenous carbon source. When such cells, suspended in phosphate buffer, were irradiated with blue light, the carbohydrates decreased faster than in darkness,[36] while the CO_2 output and the O_2 uptake increased slowly by 100% or more over that of a dark control.[37-39] From 5 to 10 min elapsed before the enhanced rate was stabilized. The intensity dependence of the loss in carbohydrate revealed saturation at the above fluence rate of about 300 $\mu W \cdot cm^{-2}$,[36] and that of the increase in O_2 uptake in some cases at even lower fluence rates.[40-42] But this difference probably depends on methodical reasons only. The wavelength dependences in both cases exhibited the above maxima around 370 nm and at 460 to 480 nm.[36,40-42] After turning the light off, the enhanced O_2 uptake decreased slowly reaching the low rate of a dark control after only several hours.

B. With Concurrent Photosynthesis

When blue light is turned on, the photosynthetic O_2 liberation of *Chlorella* is greater in the beginning than after 10 min of irradiation. During this period of time the initially high O_2 output decreases gradually to a lower, permanent rate. Such a drop does not occur in red light, in which — very short-lived induction phenomena excluded — the permanent rate is adjusted immediately. Since the measurable O_2 exchange is always the result of photosynthetic O_2 production and respiratory O_2 consumption, the above effect might result from a gradually increasing blue-light-dependent uptake of oxygen. This idea is supported by the resemblance of both time courses, i.e., induction periods of 5 to 10 min. For further support, a wavelength dependence of the gradual decline of O_2 liberation has been determined.[43] It shows pronounced maxima around 470 nm and in the near UV and thus matches the wavelength dependence of the light-enhanced respiratory O_2 consumption in DCMU-poisoned or chlorophyll-free mutated cells. Beside this there is a second hint for an enhancement of respiration during photosynthesis: in darkness following a photosynthetic light period, the O_2 uptake is greater at the beginning, and declines within several minutes to a more-or-less permanent lower rate. This enhanced dark-O_2 uptake is greater after blue light than after red light, and it also declines more slowly after blue irradiation. These results were first described by Emerson and Lewis in 1943[44] and were later confirmed by Kowallik and Kowallik.[43] The latter authors tried to determine a wavelength dependence for the extent of the enhanced dark-O_2 uptake following a light period of various colors. Based on the O_2 uptake after red light, the approximately calculated spectrum for additionally consumed O_2 revealed a steep maximum around 470 nm and also activity in the near UV. This again matches the wavelength dependences for light-enhanced respiratory oxygen uptake described above. Independently, the same result has been reported by French's laboratory.[45-47] When *Chlorella* cells were illuminated with flashes of different colors of light, the O_2 uptake thereafter was greater after a blue than after a red flash. In addition the enhanced rate decreased rather rapidly after red, but only slowly in the course of about 10 min after blue light.[45,46] The wavelength dependence of this effect showed the above maxima in the blue and near UV, too.[47]

In conclusion, there is good evidence for an enhancement of respiration by blue light in nonphotosynthesizing plant cells and some indication for such an event during photosynthesis, too.

V. INFLUENCES OF BLUE LIGHT ON ENZYME ACTIVITIES

What is the point of attack of blue light on the oxidative degradation of the carbohydrates of the cell? The RQ of unity and the results of feeding experiments with exogenous radioactive glucose revealing identical metabolites in darkness and in blue light[36] led to the conclusion that it is the mitochondrial dark type of respiration which is enhanced by short-wavelength visible radiation. Also, the results of poisoning experiments with cyanide support this conclusion.[48] In addition, α-hydroxymethane sulfonate and high oxygen pressures did not modify the effect,[49] which speaks against an involvement of photorespiration, i.e., the formation of glycolate from ribulose-1,5-bisphosphate and its oxidation to glyoxylate with further synthesis of glycine and serine. In trying to pin down the effect, it was found that anaerobic degradation of reserve carbohydrates is also enhanced by blue light.[50] Intensity and wavelength dependences match those for respiration,[51] which localizes the light effect within the glycolytic part of carbohydrate degradation. Therefore, events have been looked for which speed up that metabolic pathway.

Glycolysis is regulated especially by phosphofructokinase and pyruvate kinase, two enzymes with low capacities, which makes them "bottlenecks" for the whole pathway. The in vitro activity of phosphofructokinase is scarcely affected by blue irradiation of the living cell,[52] while that of pyruvate kinase was found to be increased by up to 100%.[53] This increase by illumination showed intensity and wavelength dependences like those of the enhanced loss in carbohydrate, O_2 uptake and fermentation, making a connection of all these events likely.[54] Inhibitor studies with cycloheximide and actinomycin point to an increase in the total amount of enzyme; K_m determinations in addition led to the assumption that there are two species of pyruvate kinase, the one with greater affinity to its substrate being increased on blue light irradiation[55,56] (see also Chapter 5 in Volume 2). Further studies on the other enzymes catalyzing carbohydrate breakdown revealed only minor effects on most of them (Chapter 5, Volume 2); they led, however, to the observation of an inhibition of α-ketoglutarate dehydrogenase. This tricarboxylic acid cycle enzyme oxydizes α-ketoglutarate (= 2-oxoglutarate) to succinyl-CoA. Again intensity and wavelength dependences correspond to those of the increased respiratory parameters and the activity of pyruvate kinase.[57] Inhibition of one component of an increased metabolic pathway is contradictory on a first glimpse. It might, however, be considered a useful adaptation, in case its substrate is used for some other reaction. α-Ketoglutarate is also the substrate for glutamate-producing enzymes. Indeed, one of them, glutamine-2-oxoglutarate-aminotransferase, was found to be increased by illumination of the yellow *Chlorella* mutant cell by up to 100%. Again, intensity and wavelength dependences look like those of the carbohydrate catabolism.[58] From these data there appears to be an overall regulation by blue light of the metabolism of the cell, finally leading to an enhanced accumulation of protein.

VI. HYPOTHESES ON THE MECHANISM OF BLUE LIGHT ACTION

The above increase in the capacity of a "bottleneck" enzyme by a light-dependent enhancement of the amount and affinity of the enzyme does not suffice to explain the observed enhancement of respiration since the extent and the dependence on the time of illumination are very different. While the increase in O_2 uptake can amount to up to ten times, already reaching a permanent rate after 5 to 10 min,[37] the enhancement in enzyme capacity develops within 12 to 15 hr and at most amounts to 100%.[54] In subsequent darkness the increased enzyme capacity remains high for several hours,[59] while the enhanced O_2 uptake begins to drop immediately after turning the light off.[37] Thus, beside the two effects mentioned in the previous chapter, a third effect of blue light on enzyme activities has to be considered.[60] This fast response might depend on alterations in substrate and/or cofactor concentrations at the site of enzyme actions, i.e., it could affect its fine regulation. One, therefore, might have to consider more than one single mechanism of the blue light action. Hypothetically two different events are being discussed which may be called direct and indirect effects.

A. Direct Effects

If an enzyme contains a light-absorbing component constitutively, an effect on its activity by modifying charge and structure of the protein by excitation of the chromophoric group can easily be seen. From the known chromophoric groups in the case described here, such a component might be a flavin or a carotenoid. But for neither phosphofructokinase or for pyruvate kinase are such pigments known to be an integral part. Among the respiratory enzymes, only succinate dehydrogenase contains a flavin as the prosthetic group. In vitro experiments with isolated mitochondria have revealed an inhibition,[61] while with enzyme preparation of various sources no effect of blue light on its activity has been detected.[61a] For other nonrespiratory, flavin-containing enzymes like glycolate oxidase[62,63] and nitrate reductase,[64] however, influences of blue irradiation have been described. However, non-flavin-containing enzymes have also been reported to respond to that radiation when flavins were added in vitro. Thus glycine oxidase has been described to increase, while 1-lactate dehydrogenase, urate oxidase, xanthin oxidase, malate dehydrogenase, transketolase, and ribulose-1,5-bisphosphate carboxylase all were found to be inhibited by blue light in the presence of exogenous flavin.[65-69] From this, an intervention of blue light absorbing pigments with chromophoreless enzyme molecules must also be considered to occur in vivo.

B. Indirect Effects

Alterations of substrate and/or cofactor distribution in different cell compartments might regulate the enzyme activity in vivo as well as *de novo* synthesis via substrate induction. Unfortunately, no pool size measurements are available at present. Therefore, the matter can only be treated purely hypothetically. Flavins as well as carotenoids might be incorporated in or be attached to compartmentalizing membranes, modifying their permeability by influences on the structure of neighboring proteins. Flavins might do this from the excited state, and carotenoids by alteration from the *cis-* to the *trans-* form. While we have to wait for direct experimental results on this problem, indirect hints do already exist. First, addition of glucose to the medium results in a greater pyruvate kinase activity, which inhibitor studies show to be caused partly by *de novo* synthesis.[53] This might indicate substrate induction of the biosynthesis of the enzyme. Second, Laudenbach and Pirson[16] have reported on a specific fraction of carbohydrate differently accumulated in *Chlorella* cells from red or blue light autotrophic cultures. Localizing this material within the chloroplast, the authors assume that the enhanced degradation of carbohydrate under blue light depends on a greater flow of substrate from the storage compartment. Third, on poisoning of glyceraldehyde-3-phosphate dehydrogenase with monoiodoacetate, the amount of water-extractable sugar and sugar phosphates from *Chlorella* was clearly enhanced after blue light exposure.[53] Perhaps this indicates an increased transfer of metabolites out of the carbohydrate-storing compartment — the chloropast — into the cytosol. Fourth, and finally, the temperature dependence of the uptake of 3-O-methylglucose from the medium of chlorophyll-free *Chlorella* cells matches that for increased O_2 uptake in blue light, which is clearly different from that of dark-O_2 uptake.[70] This also allows speculations on a blue light effect on the membranes of the cell.

VII. SIGNIFICANCE OF THE EFFECTS OF BLUE LIGHT

The only hypothesis discussed thus far proposes a connection between the enhanced carbohydrate degradation and the formation of the photosynthetic apparatus. This idea derives from transition experiments with various algae from heterotrophic to autotrophic conditions. The light-dependent greening of etiolated *Euglena*[71] and that of suitably mutated *Scenedesmus*[72] has been examined most intensively. In both cases the production of chlorophyll and the chloroplast fine structure follow an action spectrum matching that of enhanced respiration, which latter was always found to accompany greening. This topic is treated more closely in Chapter 7 of this volume.

Besides the formation of the photosynthetic apparatus, however, its maintenance also depends on blue radiation. In pure red light, the highly organized thylakoid membranes become disorganized and the photosynthetic capacity decreases.[73-77] This damage can be healed by application of trace amounts of blue light, the action spectrum for the repair exhibiting maxima at 460 and 370 nm,[73,78] thus matching those for the enhancement of respiration. Indeed, an increase in respiration has been seen to precede the restoration of photosynthesis of *Ankistrodesmus* imparied by prolonged red light exposure, too.[74] There are two more influences of blue light which might be indicative of its necessity for the optimization of the photosynthetic process. First, there is O. Warburg's old report on a so called "catalytic" effect of blue-green light, meaning an improving influence by that radiation on the photosynthetic quantum efficiency.[79,80] Second, there are reports on the formation of an unusual chlorophyll a/b ratio of about 5:1 in red light which can be changed to that of 3:1 of white light cultures by the addition of very small amounts of blue light. Again, this light action follows the above wavelength dependences.[81,82] Since the major part of chlorophyll b is thought to be incorporated in the pigment-protein complexes of photosynthetic photosystem II, a ratio of 3:1 might be indicative of an appropriate balance of photosystems I and II, usually adjusted in white light by the action of its blue quanta.

VIII. SUMMARY

Summing up, there is an effect of blue light on the level of reserve carbohydrate and of protein in plant cells. Under the influence of such radiation, the amount of carbohydrate is kept lower and that of protein higher than in darkness or under red light. Respiration is increased, and this enhancement is accompanied by (1) *de novo* synthesis of "bottleneck" enzymes, (2) production of enzyme species with greater substrate affinity, and perhaps (3) an altered distribution of substrate and cofactors within various cell compartments. The latter might regulate enzyme activities by fine regulation; it might, however, also be responsible for the observed coarse regulation. The significance of this effect of blue light is not clear yet; morphogenetic effects are being discussed.

ACKNOWLEDGMENT

The author thanks Miss P. Gayk and Mr. R. Church for their support.

REFERENCES

1. **Voskresenskaya, N. P.,** Influence of the spectral distribution of light on the products of photosynthesis, *Dokl. Akad. Nauk SSSR,* 86, 429, 1952.
2. **Voskresenskaya, N. P.,** The significance of the spectral composition of light in photosynthetic formation of substances, *Dokl. Akad. Nauk SSSR,* 93, 911, 1953.
3. **Appleman, D. and Pyfrom, H. T.,** Changes in catalase activity and other responses induced in plants by red and blue light, *Plant Physiol.,* 30, 543, 1955.
4. **Howell, R., Krober, O., and Collins, F.,** The effect of light quality on growth and composition of soybeans, *Plant Physiol.,* 32(Suppl.), 8, 1957.

5. **Casper, R. and Pirson, A.,** Grund- und Säurestoffwechsel in Blät-tern von *Kalanchoe rotundifolia* Haw. bei Farblichtkultur, *Flora,* 15b, 177, 1965.
6. **Ohlenroth, K. and Mohr, H.,** Die Steuerung der Proteinsynthese und der Morphogenese bei Farnvorkeimen durch Licht, *Planta,* 59, 427, 1963.
7. **Ohlenroth, K. and Mohr, H.,** Die Steuerung der Proteinsynthese durch Blaulicht und Hellrot in den Vorkeimen von *Dryopteris filix-mas* (L.) Schott, *Planta,* 62, 160, 1964.
8. **Bergfeld, R.,** Die Wirkung von hellroter und blauer Strahlung auf die Chloroplastenbildung, *Z. Naturforsch. Teil B,* 18, 328, 1963.
9. **Bergfeld, R.,** Der Einfluss roter und blauer Strahlung auf die Ausbildung der Chloroplasten bei gehemmter Proteinsynthese, *Z. Naturforsch., Teil B,* 1076, 1964.
10. **Deimling, A. v. and Mohr, H.,** Eine Analyse der durch Blaulicht bewirkten Steigerung der Proteinsynthese bei Farnvorkeimen auf der Ebene der Aminosäuren, *Planta,* 76, 269, 1967.
11. **Raghavan, V. and Demaggio, A. E.,** Enhancement of protein synthesis in isolated chloroplasts by irradiation of fern gametophytes with blue light, *Plant Physiol.,* 48, 82, 1971.
12. **Pirson, A. and Kowallik, W.,** Wirkung des blauen und roten Spektralbereiches auf die Zusammensetzung von *Chlorella* bei Anzucht im Licht-Dunkel-Wechsel, *Naturwissenschaften,* 47, 476, 1960.
13. **Kowallik, W.,** Über die Wirkung des blauen und roten Spektralbereichs auf die Zusammensetzung und Zellteilung synchronisierter Chlorellen, *Planta,* 58, 337, 1962.
14. **Kowallik, W.,** Die Proteinproduktion von *Chlorella* im Licht verschiedener Wellenlängen, *Planta,* 64, 191, 1965.
15. **Buschbom, A.,** Vergleichend-physiologische Untersuchungen an Algen aus Blau- und Rotlichtkulturen, *Ph.D. thesis,* University of Göttingen, Göttingen, West Germany, 1968.
16. **Laudenbach, B. and Pirson, A.,** Über den Kohlenhydratumsatz in *Chlorella* unter dem Einfluss von blauem und rotem Licht, *Arch. Mikrobiol.,* 67, 226, 1969.
17. **Andersag, R. and Pirson, A.,** Verwertung von Glucose in *Chlorella*-kulturen bei Blau- und Rotlichtbestrahlung, *Biochem. Physiol. Pflanz.,* 169, 71, 1976.
18. **Stabenau, H.,** Aktivitätsänderungen von Enzymen bei *Chlorogonium elongatum* unter dem Einfluss von rotem und blauem Licht, *Z. Pflanzenphysiol.,* 67, 105, 1972.
19. **Steup, M.,** Über Beziehungen zwischen Kohlenhydrat-, Protein- und Phosphatstoffwechsel bei *Chlamydomonas reinhardii* Dangeard im blauen und roten Spektralbereich, *Ph.D. thesis,* University of Göttingen, Göttingen, West Germany, 1972.
20. **Clauss, H.,** Über den Einfluss von Rot- und Blaulicht auf das Wachstum kernhaltiger Teile von *Acetabularia mediterranea, Naturwissenschaften,* 50, 719, 1963.
21. **Clauss, H.,** Beeinflussung der Morphogenese, Substanzproduktion und Proteinzunahme von *Acetabularia mediterranea* durch sichtbare Strahlung, *Protoplasma,* 65, 49, 1968.
22. **Brinkmann, G. and Senger, H.,** The development of structure and function in chloroplasts of greening mutants of *Scenedesmus, Plant Cell Physiol.,* 19, 1427, 1978.
23. **Jeeji Bai, N. and Subramanian, S.,** Growth, protein and pigment synthesis in *Spirulina fusiformis* in blue light, presented at 2nd Int. Conf. Effect of Blue Light in Plants and Microorganisms, Marburg, West Germany, July 10 to 17, 1983.
24. **Cayle, T. and Emerson, R.,** Effect of wave-length on the distribution of carbon-14 in the early products of photosynthesis, *Nature,* 179, 89, 1957.
25. **Hauschild, A. H. W., Nelson, C. D., and Krotkov, G.,** The effect of light quality on the products of photosynthesis in *Chlorella vulgaris, Can. J. Bot.,* 40. 179, 1962.
26. **Hauschild, A. H. W., Nelson, C. D., and Krotkov, G.,** The effect of light quality on the products of photosynthesis in green and blue-green algae, and in photosynthetic bacteria, *Can. J. Bot.,* 40, 1619, 1962.
27. **Hauschild, A. H. W., Nelson, C. D., and Krotkov, G.,** Concurrent changes in the products and the rate of photosynthesis in *Chlorella vulgaris* in the presence of blue light, *Naturwissenschaften,* 51, 274, 1964.
28. **Hauschild, A. H. W., Nelson, C. D., and Krotkov, G.,** On the mode of action of blue light on the products of photosynthesis in *Chlorella vulgaris, Naturwissenschaften,* 52, 435, 1965.
29. **Ogasawara, N. and Miyachi, S.,** Effect of wavelength on $^{14}CO_2$-fixation in *Chlorella* cells, in *Progress in Photosynthesis Research,* Vol. 3, Metzner, H., Ed., Tübingen, West Germany, 1969, 1653.
30. **Ogasawara, N. and Miyachi, S.,** Regulation of CO_2-fixation in *Chlorella* by light of varied wavelengths and intensities, *Plant Cell Physiol.,* 11, 1, 1970.
31. **Kamiya, A. and Miyachi, S.,** Blue light-induced formation of phosphoenolpyruvate carboxylase in colorless *Chlorella* mutant cells, *Plant Cell Physiol.,* 16, 729, 1975.
32. **Ruyters, G.,** Blue light-enhanced phosphoenolpyruvate carboxylase activity in a chlorophyll-free *Chlorella* mutant, *Z. Pflanzenphysiol.,* 100, 107, 1980.
33. **Kowallik, W. and Schätzle, S.,** Enhancement of carbohydrate degradation by blue light, in *The Blue Light Syndrome,* Senger, H., Ed., Springer-Verlag, Berlin, 1980, 344.
34. **Kowallik, W.,** Der Einfluss von Licht auf die Atmung von *Chlorella* bei gehemmter Photosynthese, *Planta,* 86, 50, 1969.

35. **Kowallik, W.**, Einfluss verschiedener Lichtwellenlängen auf die Zusammensetzung von *Chlorella* in Glucosekultur bei gehemmter Photosynthese, *Planta*, 69, 292, 1966.

36. **Georgi, M.**, Über blaulichtbedingte Veränderungen im Kohlenhydrat- und Proteingehalt einer chlorophyllfreien Mutante von *Chlorella vulgaris*, *Ph.D. thesis*, University of Cologne, Cologne, West Germany, 1974.

37. **Kowallik, W.**, Chlorophyll-independent photochemistry in algae, in Energy Conversion by the Photosynthetic Apparatus, *Brookhaven Symp. Biol.*, 19, 467, 1966.

38. **Kowallik, W. and Gaffron, H.**, Respiration induced by blue light, *Planta*, 69, 92, 1966.

39. **Schmid, G. H. and Schwarze, P.**, Blue light enhanced respiration in a colorless *Chlorella* mutant, *Hoppe-Seylers Z. Physiol. Chem.*, 350, 1513, 1969.

40. **Kowallik, W.**, Action spectrum for an enhancement of endogenous respiration by light in *Chlorella*, *Plant Physiol.*, 42, 672, 1967.

41. **Feindler, U.**, Über den Einfluss von Licht auf den Atmungsgaswechsel einer farblosen *Chlorella*- Mutante. Energie- und Wellenlängenbedarf, *Staatsexamensarbeit*, University of Cologne, Cologne, West Germany, 1974.

42. **Kamiya, A. and Miyachi, S.**, Effects of blue light on respiration and carbon dioxide fixation in colorless *Chlorella* mutant cells, *Plant Cell Physiol.*, 15, 927, 1974.

43. **Kowallik, U. and Kowallik, W.**, Eine wellenlängenabhängige Atmungssteigerung während der Photosynthese von *Chlorella*, *Planta*, 84, 141, 1969.

44. **Emerson, R. and Lewis, C. M.**, The dependence of the quantum yield of *Chlorella* photosynthesis on wavelength of light, *Am. J. Bot.*, 30, 165, 1943.

45. **Ried, A.**, Transients of oxygen exchange in *Chlorella* caused by short light exposures, *Carnegie Inst. Washington Yearb.*, 64, 399, 1965.

46. **Ried, A.**, Interactions between photosynthesis and respiration in *Chlorella*. Types of transients of oxygen exchange after short light exposures, *Biochim. Biophys. Acta*, 153, 653, 1968.

47. **Pickett, J. M. and French, C. S.**, The action spectrum for blue-light-stimulated oxygen uptake in *Chlorella*, *Proc. Natl. Acad. Sci. U.S.A.*, 57, 1587, 1967.

48. **Schwarzmann, A.**, Vergleichende Untersuchungen zum lichtgesteigerten Atmungsgaswechsel und zur Glykolat-Oxidation einer chlorophyllfreien *Chlorella*- Mutante, Diplomarbeit, University of Cologne, Cologne, West Germany, 1973.

49. **Kowallik, W.**, Light stimulated respiratory gas exchange in algae and its relation to photorespiration, in *Photosynthesis and Photorespiration*, Hatch, M. D., Osmond, C. B., and Slatyer, R. O., Eds., Wiley-Interscience, New York, 1971, 514.

50. **Kowallik, W. and Gaffron, H.**, Enhancement of respiration and fermentation in algae by blue light, *Nature*, 215, 1038, 1967.

51. **Kowallik, W.**, Eine fördernde Wirkung von Blaulicht auf die Säureproduktion anaerob gehaltener Chlorellen, *Planta*, 87, 372, 1969.

52. **Angele, S.**, Über die Wirkung von Blaulicht auf Kohlenhydrat abbauende Enzyme bei *Chlorella*, Ph.D. thesis, University of Bielefeld, Bielefeld, West Germany, 1981.

53. **Kowallik, W. and Ruyters, G.**, Über Aktivitätssteigerungen der Pyruvatkinase durch Blaulicht oder Glucose bei einer chlorophyll-freien *Chlorella*- Mutante, *Planta*, 128, 11, 1976.

54. **Ruyters, G. and Kowallik, W.**, Further studies of the light-mediated change in the activity of pyruvate kinase of a chlorophyll-free *Chlorella* mutant, *Z. Pflanzenphysiol.*, 96, 29, 1980.

55. **Ruyters, G.**, Blue light-effects on enzymes of the carbohydrate metabolism in *Chlorella*. I. Pyruvate-kinase, in *The Blue Light Syndrome*, Senger, H., Ed., Springer-Verlag, Berlin, 1980, 361.

56. **Ruyters, G.**, Isoenzymes of pyruvate kinase in a chlorophyll-free *Chlorella* mutant and their blue-light-mediated interconversion, *Z. Pflanzenphysiol.*, 103, 109, 1981.

57. **Stein, G. and Kowallik, W.**, Inhibition by blue light of α-ketoglutarate dehydrogenase activity in a chlorophyll-free *Chlorella* mutant, *Biochem. Physiol. Pflanz.*, 180, 163, 1985.

58. **Kowallik, W. and Neuert, G.**, Enhancement of GOGAT activity by blue light in *Chlorella*, in *Blue Light Effects in Biological Systems*, Senger, H., Ed., Springer-Verlag, Berlin, 1984, 310.

59. **Ruyters, G.**, Effects of blue light on respiration and enzyme activity in a yellow *Chlorella* mutant, in *Photosynthesis, Volume 5, Chloroplast Development*, Akoyunoglou, G., Ed., Balaban International Science Services, Philadelphia, 1981, 905.

60. **Ruyters, G.**, Effects of blue light on enzymes, in *Blue Light Effects in Biological Systems*, Senger, H., Ed., Springer-Verlag, Berlin, 1984, 283.

61. **Matile, P.**, Wirkungen des sichtbaren Lichtes auf die Atmung von Hefe, *Ber. Schweiz. Bot. Ges.*, 72, 236, 1962.

61a. **Schmid, G. H.**, personal communication, 1972.

62. **Schmid, G. H.**, The effect of blue light on glycolate oxidase of tobacco, *Hoppe-Seylers Z. Physiol. Chem.*, 350, 1035, 1969.

63. **Gnanam, A., Habib Mohamed, A., and Seetha, R.,** Comparative studies on the effect of ammonia and blue light on the regulation of photosynthetic carbon metabolism in higher plants, in *The Blue Light Syndrome,* Senger, H., Ed., Springer-Verlag, Berlin, 1980, 435.
64. **Aparicio, P. J., Roldan, J. M., and Calero, F.,** Blue light photoreactivation of nitrate reductase from green algae and higher plants, *Biochem. Biophys. Res. Commun.,* 70, 1071, 1976.
65. **Schmid, G. H.,** The effect of blue light on some flavin enzymes, *Hoppe-Seylers Z. Physiol. Chem.,* 351, 575, 1970.
66. **Schmid, G. H.,** Conformational changes caused by blue light, in *The Blue Light Syndrome,* Senger, H., Ed., Springer-Verlag, Berlin, 1980, 198.
67. **Codd, G. A.,** The photoinhibition of malate dehydrogenase, *FEBS Lett.,* 20, 211, 1972.
68. **Codd, G. A.,** The photoinactivation of tobacco transketolase in the presence of flavin mononucleotide, *Z. Naturforsch. Teil B,* 27, 701, 1972.
69. **Codd, G. A. and Stewart, R.,** Photoinactivation of ribulose bisphosphate carboxylase from green algae and cyanobacteria, *FEMS Microbiol. Lett.,* 8, 237, 1980.
70. **Kowallik, W. and Kirst, R.,** Über unterschiedliche Temperaturabhängigkeiten des Atmungsgaswechsels einer chlorophyll-freien *Chlorella-* Mutante im Dunkel und im Licht, *Planta,* 124, 261, 1975.
71. **Schiff, J. A.,** Blue light and the photocontrol of chloroplast development in *Euglena,* in *The Blue Light Syndrome,* Senger, H., Ed., Springer-Verlag, Berlin, 1980, 495.
72. **Brinkmann, G. and Senger, H.,** Blue light regulation of chloroplast development in *Scenedesmus* mutant C-2A′, in *The Blue Light Syndrome,* Senger, H., Ed., Springer-Verlag, Berlin, 1980, 526.
73. **Terborgh, J.,** Potentiation of photosynthetic oxygen evolution in red light by small quantities of monochromatic blue light, *Plant Physiol.,* 41, 1401, 1966.
74. **Adler, K.,** Spezifische Rolle der Carotinoidabsorption bei der photosynthetischen Sauerstoffentwicklung, *Planta,* 75, 220, 1967.
75. **Schael, U. and Clauss, H.,** Die Wirkung von Rotlicht und Blaulicht auf die Photosynthese von *Acetabularia mediterranea, Planta,* 78, 98, 1968.
76. **Voskresenskaya, N. P., Nechaeva, E. P., Vlasova, M. P., and Nichiporovich, A. A.,** Significance of blue light and kinetin for restoration of the photosynthetic apparatus of aging barley leaves, *Fiziol. Rast.,* 15, 890, 1968.
77. **Voskresenskaya, N. P.,** Blue light and carbon metabolism, *Annu. Rev. Plant Physiol.,* 23, 219, 1972.
78. **Ried, A.,** Über die Wirkung blauen Lichtes auf den photosynthetischen O_2-Austausch von *Chlorella, Planta,* 87, 333, 1969.
79. **Warburg, O., Krippahl, G., and Schröder, W.,** Katalytische Wirkung des blaugrünen Lichts auf den Energieumsatz bei der Photosynthese, *Z. Naturforsch. Teil B,* 9, 667, 1954.
80. **Warburg, O., Krippahl, G., and Schröder, W.,** Wirkungsspektrum eines Photosynthese-Ferments, *Z. Naturforsch. Teil B,* 10, 631, 1955.
81. **Humbeck, K., Schumann, R., and Senger, H.,** The influence of blue light on the formation of chlorophyll-protein complexes in *Scenedesmus,* in *Blue Light Effects in Biological Systems,* Senger, H., Ed., Springer-Verlag, Berlin, 1984, 359.
82. **Kowallik, W. and Schürmann, R.,** Chlorophyll a/b ratios of *Chlorella vulgaris* in blue or red light, in *Blue Light Effects in Biological Systems,* Senger, H., Ed., Springer-Verlag, Berlin, 1984, 352.

Chapter 3

PHOTOREGULATION OF EUKARYOTIC NITRATE REDUCTASE

Helga Ninnemann

TABLE OF CONTENTS

I. INTRODUCTION

Our concept that, parallel to their catalytic function, "appropriate" enzymes might be able to act as photoreceptors for a number of light-regulated biological processes offers an alternative to the common view of specialized molecules with an exclusive light-absorbing function as actual photoreceptors.

When Butler[1,2] and co-workers discovered blue light-induced absorbance changes in mycelia of *Phycomyces* and *Neurospora* (and in plasmodia of *Dictyostelium*), a first handle became available to screen the transition of a photophysical act into a biochemical reaction in vivo. An action spectrum revealed a flavin or flavoprotein as photoreceptor which mediates the photoreduction of a b-type cytochrome seen as absorbance increases at around 423 and 557 nm. Butler et al.[1,2] regarded the light-induced redox change as a step close to the primary act in the photocontrol mechanism and used the absorbance change as an assay system for a primary process in the vicinity of the photoreceptor. Since this discovery, flavohemoproteins with their closely adjacent prosthetic groups have gained special interest among photobiologists.

In the presence of an electron donor for excited flavin such as EDTA or certain amino acids,[3,4] principally any flavin-cytochrome couple can be photoreduced in vitro (e.g., flavin-cytochrome solutions,[5] isolated respiratory enzyme complexes I + III,[6] isolated yeast lactate dehydrogenase [cytochrome b_2], and isolated nitrate reductase of *Neurospora*[7,8]), and also in vivo (e.g., cytochromes in mitochondria isolated by osmotic shock from protoplasts of fungi and higher plants without exogenous flavin[9]), provided they are oxidized enough in the cell. For a cytochrome b specific for plasma membrane of corn, a selective photoreduction with red light in the presence of methylene blue at 426[10] or 428 nm[11] was claimed; the same was proposed for a cytochrome b of a 20,000-g pellet of *Neurospora* irradiated with blue light.[12]

Judging from action spectra, many blue light-controlled biological responses are mediated by flavin or flavoprotein photoreceptors;[13,14] also carotenoids are discussed for some blue light reactions.[15,16] From a conceptual point of view and from comparison of a host of action spectra for blue light responses, there is no compelling necessity to assume one single flavin (or carotene) photoreceptor for all biological blue light reactions, especially if one extends the range of possible photoreceptors from free flavins to flavoproteins. Until we know better, we should instead analyze each blue light response separately for its individual photoreceptor.

Work in our laboratory has concentrated on looking for a correlation between light-induced absorbance changes in mycelium of the carotenoidless albino-band (al-2 bd) mutant of the fungus *N. crassa* and light-induced changes in physiological light-controlled responses. Such a correlation could not be found for light-induced changes of rhythmic conidiation;[17,18] it was observed, however, for light-promoted conidiation after starvation conditions:[17] light-induced absorbance changes at 423 and 557 nm could be seen in vivo in *Neurospora* mycelium only at a time when light stimulated conidiation. A correlation between cytochrome b deficiency and reduced sensitivity to light in the cytoplasmic *Neurospora* mutant poky was also reported.[19]

There are conspicuous parallels between induction of conidiation and induction of synthesis or activity of nitrate reductase (NR) as well as between inhibition of conidiation and suppression of NR. In many fungi, nitrate stimulates conidiation.[20] Conidiation of *Neurospora* is maximally supported by nitrate as sole nitrogen source in the medium, whereas the mycelium grows optimally with ammonium nitrate or diammonium citrate.[21,22] On diammonium citrate (+2% sucrose) conidiation was suppressed.[22] Ninnemann[9] and Klemm-Wolfgramm[23] found that hyphae grown with nitrate formed conidia under starvation conditions (i.e., with both nitrogen and carbon sources depleted), whereas conidia occurred much later or not at all when the mycelium was starved after growth in medium containing NH_4NO_3 or NH_4Cl.

Nitrate reductase is primarily induced by nitrate and suppressed/inhibited by ammonium. The same amino acids (e.g., tryptophan, alanine, histidine) appear to be favorable for conidiation[24] and induction of NR.[25] We have obtained preliminary results showing that the light requirement for promotion of conidiation was higher in NR-deficient *nit* mutants of *Neurospora* than in wild-type or albino-band with normal NR;[26] it decreased again in somatic fusion products nit-1 + 3 (made from protoplasts of *nit*-1 and *nit*-3 mutants and selected on nitrate medium) as compared with the parental strains.[27,28]

We therefore proposed NR as a photoreceptor candidate for blue-light-stimulated conidiation in starved *Neurospora* mycelium.[29] A number of properties predestine this enzyme as a transmitter of environmental signals: its property of being chemically regulated, its changes of activity and structure under nutritional stress,[29] and the diversity of its prosthetic groups including flavin and a cytochrome b and a molybdopterin[9,69] which open it up to photoregulation so that NR-mediated light signals could be expressed as a change in a physiological response.

Here records of photoregulation of eukaryotic nitrate reductases will be summarized and discussed. A review on modulation of biological activities of NR by blue light and reversible absorbance changes has been published previously.[30]

II. EUKARYOTIC NITRATE REDUCTASE

A. Location of Nitrate Reductase

Nitrate reductases are fairly large and complicated molecules mediating the two-electron step from nitrate to nitrite. The natural electron donor in the cell is NAD(P)H with nitrate acting as acceptor (i.e., "total activity"); upon the addition of cytochrome c, ferricyanide, or dichlorphenol indophenol, the partial NAD(P)H-cytochrome c (or diaphorase) activity is seen; with artificial donors such as reduced methyl viologen (MVH) or reduced FAD, terminal or MVH-NO_3^- activity can be assayed.

The location of NR within the cell is still disputed: upon cell homogenization, the enzyme behaves like a soluble protein with the majority of its activity recovered in the supernatant of a 20 min/20,000 g centrifugation, and at best a small part of the activity associated with the membraneous fraction. Data of Pateman and Kinghorn[31] from experiments with *Aspergillus nidulans* led to the conclusion that at least in this fungus nitrate uptake was linked to NR activity, and no nitrate transport occurred independent of nitrate reduction.

If nitrate uptake is associated with NR activity, the enzyme should be located in the plasma membrane. In electron micrographs of cell sections of *Neurospora*, Roldán et al.[32] found preferential location of ferritin-labeled NR or part of NR in the cell wall/plasma membrane region and in the tonoplast after they had treated the sections with purified rabbit anti-NR IgG and labeled it with ferritin-conjugated goat antirabbit IgG. Dalling et al.,[33] however, observed that NR of spinach and tobacco leaves was located in the cytosol but bound indiscriminately to membrane (protein) surfaces unless the extraction medium was fortified with exogenous protein.

Evidence was reported for NR of green organisms that the enzyme was attached to chloroplast membranes[34] or loosely bound to microsomes or microbody-like organelles which may attach to chloroplasts after irradiation.[35] Generally, microbodies seem to be attached to chloroplasts in vivo[36,37] and detach from them under defined conditions (darkness of changing ionic strength).[38] Part of NR from homogenates of barley roots was found to be membrane (mitochondria) associated.[39] Some authors claim this membrane-bound form of cereal root NR was due to bacterial contamination.[40] Butz and Jackson[41] proposed a membrane-bound enzyme with nitrate-transporting, nitrate-reducing, and ATPase activities in order to explain the correlation between nitrate uptake and nitrate reduction often seen in higher plants. In bacteria, membrane-bound NR has been found by van't Riet and Planta.[42] Many scientists, however, consider NR to be entirely soluble (for a review, see Lee[43]).

Table 1
COMPARISON OF NITRATE REDUCTASES FROM DIFFERENT ORGANISMS

Organism	Mol wt of NR	No. of subunits	Heteromeric	IEP	Ref.
Hansenula anomala	215,000	4?			47
Neurospora crassa	230,000	2	No		48
	272,000	2	No		49
	290,000	2	No		50
				5.15	50a
Aspergillus nidulans	180,000	2	Yes		51
		3	Yes		52
				6.12	53
Chlorella vulgaris	360,000	4		4.5	54
Chlamydomonas reinhardii	220,000	4	Yes		55
				5.05[a]	56
Ankistrodesmus braunii	467,400	8	No		57
Spinacea oleracea	197,000	2 + 2?	Yes		58
				3.5 + 4.9	59
Nicotiana tabacum	200,000				60

[a]   Isoelectric point determined for NR diaphorase moiety.

The discussion of soluble and membrane-associated NR is important with respect to the possible photoreceptor role of the enzyme: a (at least transient) membrane association and an ordered orientation would facilitate the understanding for the specificity of a photoreaction and the limited range of action of a given photoreceptor for a certain photoresponse.

B. Structure of Nitrate Reductase

Comparison of nitrate reductases of eukaryotic cells reveals a number of common features, although differences between NR from different tissues and organisms (e.g., with respect to constitutivity, regulation of activity, dissociation of FAD, and protection of NAD(P)H-dependent activities[44,45]) should not be underestimated. The molecular weights of eukaryotic nitrate reductases range between 180,000 and about 500,000 (see also Hewitt and Notton[46]); isoelectric points as far as they have been published lie in comparable acidic pH ranges (see Table 1, with references 47 to 60). All nitrate reductases are sensitive against p-hydroxy-mercury benzoate (SH-center affected) and cyanide, thiocyanate, and azide (molybdenum cofactor affected[61]). The early findings that *Neurospora* NR contains FAD,[62] molybdenum,[63] and cytochrome b-557[64] with a midpoint potential at around -60 mV (NR of spinach[65]) or -73 mV (NR of *Ankistrodesmus braunii*[66]) are valid for all eukaryotic nitrate reductases isolated so far. The probable sequence of these prosthetic groups in the NR molecule is represented in Scheme 1:

$$FADH_2 \qquad MVH$$

$$NAD(P)H \longrightarrow (SH \longrightarrow FAD \longrightarrow Cyt.\ b\text{-}557)_n \longrightarrow Pter\text{-}Moco \qquad 2e^-$$

Cyt. c, DCPIP, ferricyanide   ferricyanide

———— diaphorase activity ————   ———— terminal activity ————

———————————— total activity ————————————

Recent studies of the Mo cofactor of NR from *Chlorella* and *Neurospora* revealed a pterin associated with the cofactor which is seen fluorometrically when the cofactor is degraded oxidatively.[9,67-69] This molybdopterin then can be excited with 410 nm[9,69] (or 380 nm[68]) and is maximally emitting at 480 nm[9,69] (or 470 nm[68]).

These different chromophoric groups render this enzyme suitable for photoregulation, possibly at different sections of the molecule. The redox states of the Mo cofactor, of FAD, and of cytochrome b-557 can be considered as points of control allowing reversible (re-)activation or inactivation. Photoregulation has in fact been observed in NR in vivo and in vitro. A relationship between light and nitrate reductase was already established when Warburg[70] showed that light stimulated nitrate reduction in *Chlorella*. Since then the generality of this observation has been demonstrated in lower and higher plants.

III. PHOTOREGULATION OF NITRATE REDUCTASE

A. Photoregulation In Vivo
1. Photoactivation and Photoinduction

Photoregulation of NR happens at two levels: by photoinduction of its synthesis or by photoactivation of already existing molecules. Both kinds of photocontrol of NR have been found by several authors. In addition, an indirectly enhanced NR activity is suspected as a consequence of stimulated nitrate uptake.

Burström[71] concluded that wheat leaves reduced nitrate in light and not in darkness. Total and specific activity of extractable NR decreased in leaves of cauliflower grown in darkness, whereas both remained high in light or, once they had decreased after transfer of plants from light to dark (dark-adaptation), increased again in the light to the level at the beginning of the light-dark transfer.[72] Also Hageman and Flesher[73] found an increase of maize NR after illumination of leaves and lower activities in shaded leaves or in those transferred into darkness. Here again the NR activity level was restored after subsequent irradiation. The loss of enzyme activity in darkness appeared in vivo and was not an artifact during extraction and assay of the plant material. Many similar observations for higher plants and their seedlings have been reported in the literature[74,75] (for a review, see References 45 and 76).

a. Photoactivation of Nitrate Reductase via Stimulation of Nitrate Uptake

Beevers et al.[75] concluded that the enhancing effect of short-term light on NR activity in higher plants was indirect and attributed it to increased membrane permeability of the cells and thus enhanced nitrate uptake. Light-promoted availability or permeation of nitrate was also proposed by other scientists.[77-80] Also for *Chlamydomonas*, a light-stimulated NR activity was described which probably resulted from stimulated nitrate uptake.[81]

b. Direct Photostimulation of the Nitrate Reductase Molecule

In many plant systems the control mechanism operates at the level of the NR molecule. Rufty et al.[82] measured similar rates of nitrate uptake for leaves of soybean seedlings in darkness and white light, but the nitrate absorbed in darkness was reduced less efficiently than that absorbed in the light. Only a small portion of total nitrate reduction in the light occurred in the root system *in situ*. In other instances, however, a higher NR activity in light than in darkness was recorded also in roots.[83] Stoy[84,85] reported a blue light-induced increase of nitrate assimilation in wheat leaves and a riboflavin-catalyzed enzymatic photoreduction of nitrate. Tischner and Hüttermann[86] showed photoactivation of NR in synchronized *Chlorella* which was not due to *de novo* synthesis; rather, light supposedly acted directly on the NR molecule or stimulated nitrate uptake. Johnson[87] described a rapid phytochrome-controlled increase of NR activity in *Sinapis alba* (measured by an in vivo method) which was insensitive to inhibitors of protein synthesis. He suggested that light activated the existing NR, as did a number of other authors.[88,89]

c. Photostimulation of Nitrate Reductase Synthesis

In several cases an induction of NR synthesis in the light is discussed. Enhancement of NR synthesis in etiolated corn leaves and pea seedlings by blue light was observed.[90,91] In some higher plants light-stimulated *de novo* synthesis of NR appeared to be caused preferentially by red light via the phytochrome system.[77] Photocontrol of ribosomal activity via the phytochrome system was shown by Pine and Klein.[92] Also, results from experiments with inhibitors of mRNA synthesis and protein synthesis in *Ankistrodesmus* provided evidence that light caused *de novo* synthesis of NR.[93] Besides an increased capacity for protein synthesis in illuminated leaves, an increased polysome development appears to cause stimulation of NR synthesis in light; also, here red light was more effective than blue.[94-96] Decrease of NR activity in darkness was thought to be a consequence of NR decay.[76]

As an additional or alternative mechanism, a requirement for a functional photosynthetic system providing a suitable redox potential in the cytoplasm for the light-stimulated induction of NR with nitrate was claimed.[97] In addition to light and nitrate, the presence of CO_2 was found necessary for the formation of NR.[98,99] Such a dependence of photoinduction of NR on photosynthetic electron flow and the NAD/NADH ratio was, however, denied for etiolated barley shoots.[100]

2. Photoreactivation of Inactive Nitrate Reductase

NR of some algae and higher plants can exist in two metabolically interconvertible forms, an active and an inactive one, depending on the redox state of regulatory site(s) within the enzyme, the reduction of which results in enzyme inactivation. Darkness or ammonium ions decreased in vivo NR activity of *Chlorella fusca, C. vulgaris, Chlamydomonas reinhardii* and spinach; reactivation of the enzyme in intact cells occurred after removal of the ammonium ions or after addition of ferricyanide.[101-105] In *Chlamydomonas reinhardii*, NR which had turned into the inactive form during a dark period could be reactivated *in situ* by white or blue but not by red light in the presence of nitrate.[81,106]

In *Neurospora*, metabolically interconvertible forms of NR are not certain, although the fungal NR can be inactivated in vivo by addition of ammonium to the medium or by starvation conditions (for *N. crassa* wild type, see References 107 and 108; for *Neurospora* mutant al-2 bd, see References 9 and 9a). Several authors reported enzyme degradation, i.e., rapid decrease of total, terminal, and diaphorase activities, instead of reversible inactivation.[107,108] Ammonium-inactivated *Neurospora* NR from mutant albino band or wild type could not be photoreactivated by us (unpublished) or other scientists. But evidence is accumulating that NR in partially starved mycelium (after growth in nitrate medium starvation without nitrogen source) which had lost an essential percentage of total NR activity (but not of terminal activity; this is in contrast to the observation of Klein Amy and Garrett[108] for *Neurospora* wild type!) can be photoreactivated in vivo (see References 29 and 29a). Detailed preconditions for this photoreactivation have been analyzed by us.

B. Photoregulation In Vitro

Partially purified NR of various sources (algae or higher plants) can be inactivated in presence of a reductant such as NAD(P)H, dithionite, thiols.[109,110] Nitrate reductases of some species — among them *Neurospora* NR — require in addition cyanide.[111-114] Vennesland et al.[104] studied inactivation of *Chlorella* NR and found endogenous, bound cyanide in the naturally inactivated form of the enzyme.

This reductive inactivation of NR with cyanide affects the molybdenum cofactor, not the diaphorase activity.[110] Coughlan et al.[61] proposed as the mechanism for the reversible inactivation of molybdenum-containing enzymes by cyanide plus reductant an unusual state of reduction ("overreduction") of the molybdenum center; i.e., a reduction state of Mo(III) or Mo(II) which is below the valence states Mo(VI) to Mo(IV) attained during the catalytic cycle of the native enzymes.

The inactivation can be reversed by chemical oxidation with ferricyanide or — more slowly and only in some organisms — with oxygen or trivalent manganese ions.[115-118] For NR of *Neurospora* inactivated in vitro with reductant and cyanide, 50% reactivation with ferricyanide was accomplished at a redox potential of $+371$ mV at pH 7.2 with one electron transferred in the reaction.[119] For NR of *Ankistrodesmus*, de la Rosa et al.[120] reported a redox potential of $+230$ mV at pH 7.5 for reactivation with ferricyanide.

In vitro, photoreactivation with white or blue light of in vivo or in vitro inactivated NR from *Chlorella fusca* and spinach was first observed by Aparicio et al.[121] Addition of FAD (20 μM) greatly accelerated the conversion into the active form. It was concluded that this reactivation was a photooxidation of some component(s) of the electron-transport chain of the enzyme. A role of FAD as the light-absorbing pigment was shown subsequently for spinach NR.[122] The authors speculated that the role of external flavins could be the collection and transfer of energy required for reactivation, to the bound FAD of NR. NAD(P)H prevented the photoreactivation of NR, an effect which was explained by assuming reduced pyridine nucleotides to dissipate radiant energy by quenching excited flavin molecules by reduction and/or by assuming the elimination of the FAD photoreceptor after its reduction with NAD(P)H. The hypothesis was proposed that irradiation excited endogenous FAD directly or via exogenous flavins and that excited flavin acted as a strong oxidant for the reduced part of the enzyme which kept the NR inactive.

A detailed study of the mechanism of photoreactivation was undertaken by Roldán and Butler[113] and Fritz and Ninnemann[114] for *Neurospora* NR in partially purified extracts and in a highly purified form. The *Neurospora* enzyme which was reductively inactivated with cyanide could be reactivated with blue light. At the same time irradiation caused a partial and irreversible inactivation of the diaphorase moiety of the enzyme.

The action spectrum for photoreactivation showed that flavin was the photoreceptor chromophore.[113] A search for the molecule species causing these photoreactions showed triplet flavin to be responsible for the photoreactivation and singlet oxygen for the photoinactivation of NR. Lowering the oxygen tension enhanced NR photoreactivation; furthermore, participation of superoxide anions, hydrogen peroxide, or singlet oxygen in the reaction could be excluded.[114] Similar results were reported by Mauriño et al.[123] who found no inhibition or even sometimes stimulation of the photoreactivation of spinach NR under anaerobic conditions. The enhancing effect of anaerobiosis can be explained by the quenching of triplet flavins through oxygen. Other quenchers of flavin triplet such as EDTA, histidine, or KI also inhibited the photoreactivation process.[114] The reactions of NR in the light in presence of flavins are presented in Scheme 2 (from Reference 114):

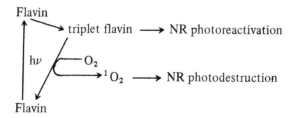

The rates of photoreactivation and photodestruction were proportional to the concentration of flavin added to the preparation; the relative effectiveness of riboflavin: FMN:FAD was 6.7:5.5:1. The bound FAD of NR remaining after enzyme purification and detectable by an increase in flavin fluorescence after SDS treatment of the purified sample proved insufficient for the photoreactivation. From the in vitro experiments, it was concluded that free flavins can act as photoreceptors in the photoreactivation and photodestruction of *Neurospora* NR. Whether bound FAD of NR participates via energy transfer from excited free flavin during a photoreactivation reaction in vivo cannot be decided from these experiments. Bound flavin of a thoroughly dialyzed preparation could, however, mediate the photoreduction of cytochrome b-557 of *Neurospora* NR, as seen from spectra monitored under anaerobic conditions; added free flavin enhanced the effect.[9]

IV. CONCLUDING REMARKS

There are fundamental differences between nitrate reductases of photosynthetic and non-photosynthetic eukaryotic cells. Therefore generalizations about regulatory reactions and mechanisms by which light influences NR activities have to be made with caution. Nitrate reductases with their large molecular weights, their long electron transport chains, and their numerous chromophoric redox groups lend themselves to chemical and photochemical regulation. Some of the effects reported in the literature may be indirect via photosynthesis and photophosphorylation, via enhanced nitrate uptake, or via a phytochrome-stimulated RNA synthesis. In several cases a direct photoregulation of NR molecules from different organisms has been proved, but in few instances has its mechanism been elucidated: so the redox state of the molybdenum cofactor determines the state of activity of the enzyme in vivo and in vitro and is accessible to chemical and to photoregulation. This photoeffect is mediated by flavin.[113] The reactive species is triplet flavin.[114] The cytochrome of the FAD-cytochrome b-557 center can be photoreduced in vivo and in vitro with blue light with or without exogenous flavin. Here the internal FAD remaining after extensive dialysis suffices to transmit the light effect (see Figure 3 in Reference 9). Whether light can influence NR activity by regulating the redox state of the FAD-cytochrome b-557 region similar to the redox mechanism exerted on the molybdenum cofactor has to be investigated. Since there are thorough homologies in the amino acid sequence of the heme-binding domains of NR, cytochrome b_2 (flavohemoprotein), and cytochrome b_5,[124] a photocontrol via the flavin-cytochrome b-557 centers may be expected in a number of enzyme activities.

A previous report[125] stated that cytochrome b-557 of *Chlorella* NR did not remain reduced after removal of reductant under conditions where the enzyme remained inactive. But the high autoxidizability of this cytochrome b, which in vitro causes the cytochrome b-557 of NR always to be seen in the oxidized form, has to be kept in mind. From this observation no conclusion can be drawn on its redox state within the cell.

Considering possible physiological responses of NR, it might be relevant that *in vitro* irradiated *Neurospora* NR can use certain amino acids or amino acid derivatives instead of EDTA as electron donors for the excited flavin to photoreduce cytochrome b-557 and to reduce nitrate to nitrite.[7,8] A photochemical reduction of nitrate has also been accomplished with NR of spinach and several microorganisms.[126]

No evidence exists so far for a participation in photocontrol of NR of the pterin associated with the molybdenum cofactor. In contrast to a pterin-like signal in the homogenate of *Neurospora* mycelium which is found already in untreated samples, the NR-specific molybdopterin signal appears only after oxidative degradation.[9,69] Its photoreceptive property in the visible or near uv region *in vivo* is uncertain. Also unknown is whether this pterin undergoes redox changes while associated with the molybdenum cofactor.

ACKNOWLEDGMENT

I gratefully acknowledge the financial support of the Deutsche Forschungsgemeinschaft for our research on *Neurospora* nitrate reductase.

REFERENCES

1. **Poff, K. L. and Butler, W. L.**, Absorbance changes induced by blue light in *Phycomyces blakesleeanus* and *Dictyostelium discoideum*, *Nature (London)*, 248, 799, 1974.
2. **Muñoz, V. and Butler, W. L.**, Photoreceptor pigment for blue light in *Neurospora crassa*, *Plant Physiol.*, 55, 421, 1975.
3. **Frisell, W. R., Chung, C. W., and Mackenzie, D. G.**, Catalysis of oxidation of nitrogen compounds by flavin coenzymes in the presence of light, *J. Biol. Chem.*, 234, 1297, 1959.
4. **Enns, E. and Burgess, W. H.**, The photochemical oxidation of ethylenediaminetetraacetic acid and methionine by riboflavin, *J. Am. Chem. Soc.*, 87, 5766, 1965.
5. **Schmidt, W. and Butler, W. L.**, Flavin-mediated photoreactions in artificial systems: a possible model for the blue-light photoreceptor pigment in living systems, *Photochem. Photobiol.*, 24, 71, 1976.
6. **Ninnemann, H., Strasser, R. J., and Butler, W. L.**, The superoxide anion as electron donor to the mitochondrial electron transport chain, *Photochem. Photobiol.*, 26, 41, 1977.
7. **Ninnemann, H. and Klemm-Wolfgramm, E.**, Blue light controlled conidiation and absorbance change in *Neurospora* are mediated by nitrate reductase, in *The Blue Light Syndrome*, Senger, H., Ed., Springer-Verlag, Berlin, 1980, 238.
8. **Ninnemann, H.**, Photoreduction of cytochrome b 557 of partially purified *Neurospora* nitrate reductase via its internal flavin, *Photochem. Photobiol.* 35, 391, 1982.
9. **Ninnemann, H.**, The nitrate reductase system, in *Blue Light Effects in Biological Systems*, Senger, H., Ed., Springer-Verlag, Berlin, 1984, 95.
9a. **Ninnemann, H.**, unpublished data, 1980.
10. **Britz, S. J., Schrott, E., Widell, S., Brain, R., and Briggs, W. R.**, Methylene blue-mediated red-light photoreduction of cytochromes in particulate fractions of corn and *Neurospora*, *Carnegie Inst. Washington, Yearb.*, 76, 289, 1977.
11. **Britz, S. J., Schrott, E., Widell, S., and Briggs, W. R.**, Red light-induced reduction of a particle-associated b-type cytochrome from corn in the presence of methylene blue, *Photochem. Photobiol.*, 29, 359, 1979.
12. **Brain, R. D., Freeberg, J. A., Weiss, C. V., and Briggs, W. R.**, Blue light-induced absorbance changes in membrane fractions from corn and *Neurospora*, *Plant Physiol.*, 59, 948, 1977.
13. **Ninnemann, H.**, Blue light photoreceptors, *BioScience*, 30, 166, 1980.
14. **Song, P. S.**, Spectroscopic and photochemical characterization of flavoproteins and carotenoproteins as blue light photoreceptors, in *The Blue Light Syndrome*, Senger, H., Ed., Springer-Verlag, Berlin, 1980, 157.
15. **De Fabo, E., Harding, R. W., and Shropshire, W.**, Action spectrum between 260 and 800 nm for the photoinduction of carotenoid biosynthesis in *Neurospora crassa*, *Plant Physiol.*, 57, 440, 1976.
16. **Harding, R. W. and Turner, R. V.**, Photoregulation of the carotenoid biosynthetic pathway in albino and white collar mutants of *Neurospora crassa*, *Plant Physiol.*, 68, 745, 1981.
17. **Klemm, E. and Ninnemann, H.**, Correlation between absorbance change and a physiological response induced by blue light in *Neurospora*, *Photochem. Photobiol.*, 28, 227, 1978.
18. **Paietta, J. and Sargent, M. L.**, Blue light responses in nitrate reductase mutants of *Neurospora crassa*, *Photochem. Photobiol.*, 35, 853, 1982.
19. **Brain, R. C., Woodward, D. O., and Briggs, W. R.**, Correlative studies of light sensitivity and cytochrome content in *Neurospora crassa*, *Carnegie Inst. Washington Yearb.*, 76, 295, 1977.
20. **Garraway, M. O. and Evans, R. C.**, Nitrogen nutrition, in *Fungal Nutrition and Physiology*, Wiley, Interscience, New York, 1984, chap. 4.
21. **Hochberg, M. L. and Sargent, M. L.**, Rhythms of enzyme activity associated with circadian conidiation in *Neurospora crassa*, *J. Bacteriol.*, 120, 1164, 1974.
22. **Weiss, B. and Turian, G.**, A study of conidiation in *Neurospora crassa*, *J. Gen. Microbiol.*, 44, 407, 1966.
23. **Klemm-Wolfgramm, E.**, Korrelation zwischen Absorptionsänderungen und physiologischer Wirkung von Blaulicht bei *Neurospora;* Nitratreduktase als möglicher Photorezeptor, Ph.D. thesis, University of Tübingen, Tübingen, West Germany, 1979.
24. **Sargent, M. L. and Kaltenborn, S. H.**, Effects of medium composition and carbon dioxide on circadian conidiation in *Neurospora*, *Plant Physiol.*, 50, 171, 1972.
25. **Subramanian, K. N., Padmanaban, G., and Sarma, P. S.**, The regulation of nitrate reductase and catalase by amino acids in *Neurospora crassa*, *Biochim. Biophys. Acta*, 151, 20, 1968.
26. **Möhren, S.**, Zur Licht-stimulierten Konidienbildung bei Neurospora-Mutanten, Master's thesis, University of Tübingen, Tübingen, West Germany, 1980.
27. **Ninnemann, H.**, Somatic hybridization of *Neurospora* protoplasts, *Plant Physiol.*, 65 (Suppl.) Abstr. 22, 1980.

28. **Ninnemann, H.,** Light-promoted conidiation in nitrate reductase-defective mutants of *Neurospora crassa* and in their nitrate reductase-complemented somatic fusion products, *Microsensory Newslett.,* 3, 16, 1981.

29. **Klemm, E. and Ninnemann, H.,** Nitrate reductase — a key enzyme in blue light-promoted conidiation and absorbance change of *Neurospora, Photochem. Photobiol.,* 29, 629, 1979.

29a. **Scheidler, L.,** Blaulichtphotorezeption und Versuche zur Lichtreaktivierung von Nitratreduktase bei *Neurospora crassa,* PhD. thesis, University of Tübingen, Tübingen, West Germany, 1986.

29b. **Ninnemann, H.,** in *Proc. Adv. Course on Inorganic Nitrogen Metabolism,* Ullrich, W., Ed., Jarandilla, Spain, 1986.

30. **Ninnemann, H.,** Reversible absorbance changes and modulation of biological activities by blue light, in *Molecular Models of Photoresponsiveness,* Montagnoli, G. and Erlanger, B. F., Eds., Plenum Press, New York, 1983, 133.

31. **Pateman, J. A. and Kinghorn, J. R.,** Nitrogen metabolism, in *the Filamentous Fungi,* Vol. 2, Smith, J. E. and Berry, D. R., Eds., J. Wiley & Sons, New York, 1976, 159.

32. **Roldán, J. M., Verbelen, J.-P., Butler, W. L., and Tokuyasu, K.,** Intracellular localization of nitrate reductase in *Neurospora crassa, Plant Physiol.,* 70, S9, 1982.

33. **Dalling, M. J., Tolbert, N. E., and Hageman, R. H.,** Intracellular location of nitrate reductase and nitrite reductase, *Biochim. Biophys. Acta,* 283, 505, 1972.

34. **Rathnam, C. K. M. and Das, V. S. R.,** Nitrate metabolism in relation to the aspartate-type C-4 pathway of photosynthesis in *Eleusine coracana, Can. J. Bot.,* 52, 2599, 1974.

35. **Lips, A. H. and Avissar, Y.,** Plant-leaf microbodies as the intracellular site of nitrate reductase and nitrite reductase, *Eur. J. Biochem.,* 29, 20, 1972.

36. **Kagan-Zur, V. and Lips, S. H.,** Studies on the intracellular location of enzymes of the photosynthetic carbon-reduction cycle, *Eur. J. Biochem.,* 59, 17, 1975.

37. **Frederick, S. E. and Newcomb, E. H.,** Microbody-like organelles in leaf cells, *Science,* 163, 1353, 1969.

38. **Lips, S. H.,** Enzyme content of plant microbodies as affected by experimental procedures, *Plant Physiol.,* 55, 598, 1975.

39. **Miflin, B. J.,** Distribution of nitrate and nitrite reductase in barley, *Nature (London),* 214, 1133, 1967.

40. **Blevins, D. G., Lowe, R. H., and Staples, L.,** Nitrate reductase in barley roots under sterile, low oxygen conditions, *Plant Physiol.,* 57, 458, 1976.

41. **Butz, R. G. and Jackson, W. A.,** A mechanism for nitrate transport and reduction, *Phytochemistry,* 16, 409, 1977.

42. **van't Riet, J. and Planta, R. J.,** Purification, structure and properties of the respiratory nitrate reductase of *Klebsiella aerogenes, Biochim. Biophys. Acta,* 379, 81, 1975.

43. **Lee, R. B.,** Sources of reductant for nitrate assimilation in nonphotosynthetic tissue: a review, *Plant Cell Environ.,* 3, 65, 1980.

44. **De la Rosa, M. A., Marquez, A. J., and Vega, J. M.,** Dissociation of FAD from the NAD(P)H-nitrate reductase complex from *Ankistrodesmus braunii* and role of flavin in catalysis, *Z. Naturforsch.,* 37c, 24, 1982.

45. **Srivastava, H. S.,** Regulation of nitrate reductase activity in higher plants, *Phytochemistry,* 19, 725, 1980.

46. **Hewitt, E. J. and Notton, B. A.,** Nitrate reductase systems in eukaryotic and prokaryotic organisms, in *Molybdenum and Molybdenum-Containing Enzymes,* Coughlan, M., Ed., Pergamon Press, 1980, 273.

47. **Zauner, E. and Dellweg, H.** Purification and properties of the assimilatory nitrate reductase from the yeast *Hansenula anomala, Eur. J. Appl. Microbiol. Biotech.,* 19, 90, 1983.

48. **Pan, S.-S. and Nason, A.,** Purification and characterization of homogenous assimilatory reduced nicotinamide adenine dinucleotide phosphate-nitrate reductase from *Neurospora crassa, Biochim. Biophys. Acta,* 523, 297, 1978.

49. **Tachiki, T. and Nason, A.,** Preparation and properties of apoenzyme of nitrate reductases from wild-type and nit-3 mutant of *Neurospora crassa, Biochim. Biophys. Acta,* 744, 16, 1983.

50. **Horner, R. D.,** Purification and comparison of nit-1 and wild-type NADPH:nitrate reductases of *Neurospora crassa, Biochim. Biophys. Acta,* 744, 7, 1983.

50a. **Pottiez, B. and Ninnemann, H.,** unpublished.

51. **Minagawa, N. and Yoshimoto, A.,** Assimilatory NADPH-nitrate reductase of *Aspergillus nidulans, J. Biochem.,* 91, 761, 1982.

52. **Downey, R. and Steiner, F. X.,** Further characterization of the NADPH-nitrate oxidoreductase in *Aspergillus nidulans, J. Bacteriol.,* 137, 105, 1979.

53. **Steiner, F. X. and Downey, R.,** Isoelectric focusing and two-dimensional analysis of purified nitrate reductase from *Aspergillus nidulans, Biochim. Biophys. Acta,* 706, 203, 1982.

54. **Howard, W. D. and Solomonson, L. P.,** Quaternary structure of assimilatory NADH:nitrate reductase from *Chlorella, J. Biol. Chem.,* 257, 10243, 1982.

55. **Franco, A. R., Cárdenas, J., and Fernández, E.,** Heteromultimeric structure of the nitrate reductase complex of *Chlamydomonas reinhardii, EMBO J.,* 3, 1403, 1984.

56. **Fernández, E. and Cárdenas, J.,** Isoelectric focusing of the NAD(P)H-cytochrome c reductase subunit of *Chlamydomonas reinhardii* nitrate reductase, *Z. Naturforsch.,* 38c, 35, 1983.

57. **De la Rosa, M. A. and Vega, J. M.,** Composition and structure of assimilatory nitrate reductase from *Ankistrodesmus braunii, J. Biol. Chem.,* 256, 5814, 1981.

58. **Notton, B. A. and Hewitt, E. J.,** Structure and properties of higher plant nitrate reductase, especially *Spinacea oleracea* (L.), in *Nitrogen Assimilation in Plants;* Hewitt, E. J. and Cutting, C. V., Eds., Academic Press, New York, 1979, 227.

59. **Notton, B. A., Hewitt, E. J., and Fielding, A. H.,** Isoelectric focusing of spinach nitrate reductase and its tungsten analogue, *Phytochemistry,* 11, 2447, 1972.

60. **Mendel, R. R. and Müller, A. J.,** Comparative characterization of nitrate reductase from wild-type and molybdenum cofactor-defective cell cultures of *Nicotiana tabacum, Plant Sci. Lett.,* 18, 277, 1980.

61. **Coughlan, M. P., Johnson, J. L., and Rajagopalan, K. V.,** Mechanism of inactivation of molybdoenzymes by cyanide, *J. Biol. Chem.,* 255, 2694, 1980.

62. **Nason, A. and Evans, H. J.,** Triphosphopyridine nucleotide-nitrate reductase in *Neurospora, J. Biol. Chem.,* 202, 655, 1953.

63. **Nicholas, D. J. D. and Nason, A.,** Molybdenum and nitrate reductase. II. Molybdenum as a constituent of nitrate reductase, *J. Biol. Chem.,* 207, 353, 1954.

64. **Garrett, R. and Nason, A.,** Involvement of a b-type cytochrome in the assimilatory nitrate reductase of *Neurospora crassa, Proc. Natl. Acad. Sci. U.S.A.,* 58, 1603, 1967.

65. **Fido, R. J., Hewitt, E. J., Notton, B. A., Jones, O. T. G., and Nasrulhaq-Boyce, Q.,** Haem of spinach nitrate reductase: low temperature spectrum and mid-point potential, *FEBS Lett.,* 99, 180, 1979.

66. **De la Rosa, M. A., Diez, J., Vega, J. M., and Losada, M.,** Purification and properties of assimilatory nitrate reductase (NAD[P]H) from *Ankistrodesmus braunii, Eur. J. Biochem.,* 106, 249, 1980.

67. **Johnson, J. L., Hainline, B. E., Rajagopalan, K. V.,** Characterization of the molybdenum cofactor of sulfite oxidase, xanthine oxidase and nitrate reductase, *J. Biol. Chem.,* 255, 1783, 1980.

68. **Johnson, J. L. and Rajagopolan, K. V.,** Structural and metabolic relationship between the molybdenum cofactor and urothione, *Proc. Natl. Acad. Sci. U.S.A.,* 79, 6856, 1982.

69. **Siefermann-Harms, D., Fritz, B., and Ninnemann, H.,** Evidence for a pterin derivative associated with the molybdenum cofactor of *Neurospora crassa* nitrate reductase, *Photochem. Photobiol.,* 42, 771, 1985.

70. **Warburg, O. and Negelein, E.,** Über die Reduktion der Salpetersäure in grünen Zellen, *Biochem. Z.,* 110, 66, 1920.

71. **Burström, H.,** Photosynthesis and assimilation of nitrate by wheat leaves, *Annu. Rev. Agric. Coll. Sweden,* 11, 1, 1943.

72. **Candela, M. I., Fisher, E. G., and Hewitt, E. J.,** Molybdenum as a plant nutrient, *Plant Physiol.,* 32, 280, 1957.

73. **Hageman, R. H. and Flesher, E.,** Nitrate reductase activity in corn seedlings as affected by light and nitrate content of nutrient median, *Plant Physiol.,* 35, 700, 1960.

74. **Sanderson, G. W. and Cocking, E. C.,** Enzymic assimilation of nitrate in tomato plants, *Plant Physiol.,* 39, 416, 1964.

75. **Beevers, L., Schrader, L. E., Flesher, D., and Hageman, R. H.,** The role of light and nitrate in the induction of nitrate reductase in radish cotyledons and maize seedlings, *Plant Physiol.,* 40, 691, 1965.

76. **Beevers, L. and Hageman, R. H.,** Nitrate and nitrite reduction, in *The Biochemistry of Plants,* Vol. 5, Miflin, B. J., Ed., Academic Press, New York, 1980, 115.

77. **Jones, R. W. and Sheard, R. W.,** Phytochrome, nitrate movement, and induction of nitrate reductase in etiolated pea terminal buds, *Plant Physiol.,* 55, 954, 1975.

78. **Rao, K. P. and Rains, D. W.,** Nitrate absorption by barley, *Plant Physiol.,* 57, 59, 1976.

79. **Tischner, R. and Lorenzen, H.,** Nitrate uptake and nitrate reduction in synchronous *Chlorella, Planta,* 146, 287, 1979.

80. **Calero, F., Ullrich, W. R., and Aparicio, P. J.,** Regulation by monochromatic light of nitrate uptake in *Chlorella fusca,* in *The Blue Light Syndrome,* Senger, H., Ed., Springer-Verlag, Berlin, 1980, 411.

81. **Florencio, F. J. and Vega, J. M.,** Regulation of the assimilation of nitrate in *Chlamydomonas reinhardii, Phytochemistry,* 21, 1195, 1982.

82. **Rufty, T. Jr., Israel, D. W., and Volk, R. J.,** Assimilation of $^{15}NO_3^-$ taken up by plants in the light and in the dark, *Plant Physiol.,* 76, 769, 1984.

83. **Deane-Drummond, C. E., Clarkson, E. T., and Johnson, C. B.,** Effect of shoot removal and malate on the activity of nitrate reductase assayed *in vivo* in barley roots, *Plant Physiol.,* 64, 660, 1979.

84. **Stoy, V.,** Action of different light qualities on simultaneous photosynthesis and nitrate assimilation in wheat leaves, *Physiol. Plant.,* 8, 963, 1955.

85. **Stoy, V.,** Riboflavin-catalyzed enzymic photoreduction of nitrate, *Biochim. Biophys. Acta,* 21, 395, 1956.

86. **Tischner, R. and Hüttermann, A.,** Light-mediated activation of nitrate reductase in synchronous *Chlorella, Plant Physiol.,* 62, 284, 1978.

87. **Johnson, C. B.,** Rapid activation by phytochrome of nitrate reductase in the cotyledons of *Sinapis alba,* *Planta,* 128, 127, 1976.
88. **Sasakawa, H. and Yamamota, Y.,** Effects of red, far red, and blue light on enhancement of nitrate reductase activity and on nitrate uptake in etiolated rice seedlings, *Plant Physiol.,* 63, 1098, 1979.
89. **Sharma, A. K. and Sopory, S. K.,** Independent effects of phytochrome and nitrate on nitrate reductase and nitrite reductase activities in maize, *Photochem. Photobiol.,* 39, 491, 1984.
90. **Jones, R. W. and Sheard, R. W.,** Effects of blue and red light on nitrate reductase level in leaves of maize and pea seedlings, *Plant Sci. Lett.,* 8, 305, 1977.
91. **Rao, L., Datta, N., Guha-Mukherjee, S., and Sopory, S. K.,** The effect of blue light on the induction of nitrate reductase in etiolated excised maize leaves, *Plant Sci. Lett.,* 28, 39, 1982/83.
92. **Pine, K. and Klein, A. O.,** Regulation of polysome formation in etiolated bean leaves by light, *Dev. Biol.,* 28, 280, 1972.
93. **Fischer, S. and Simonis, W.,** Tagesperiodische Schwankungen und lichtinduzierte Zunahme der Nitratreduktase-Aktivität bei Synchronkulturen von *Ankistrodesmus braunii, Z. Pflanzenphysiol.,* 92, 143, 1979.
94. **Williams, G. R. and Novelli, G. D.,** Stimulation of an *in vitro* amino acid incorporating system by illumination of dark-grown plants, *Biochem. Biophys. Res. Commun.,* 17, 23, 1964.
95. **Travis, R. L., Huffaker, R. C., and Key, J. L.,** Light-induced development of polyribosomes and the induction of nitrate reductase in corn leaves, *Plant Physiol.,* 46, 800, 1970.
96. **Travis, R. L. and Key, J. L.,** Correlation between polyribosome level and the ability to induce nitrate reductase in dark-grown corn seedlings, *Plant Physiol.,* 48, 617, 1971.
97. **Sawhney, S. K. and Naik, M. S.,** Role of light in the synthesis of nitrate reductase in rice seedlings, *Biochem. J.,* 130, 475, 1972.
98. **Kannangara, C. G. and Woolhouse, H. W.,** Changes in the enzyme activity of soluble protein fractions in the course of foliar senescence in *Perilla frutescens, New Phytol.,* 67, 533, 1968.
99. **Thacker, A. and Syrett, P. J.,** The assimilation of nitrate and ammonium by *Chlamydomonas reinhardii, New Phytol.,* 71, 423, 1972.
100. **Nasrulhaq-Boyce, A. and Jones, O. T. G.,** The light-induced development of nitrate reductase in etiolated barley shoots: an inhibitory effect of laevulinic acid, *Planta,* 137, 77, 1977.
101. **Losada, M., Paneque, A., Aparicio, P. J., Vega, J. M., Cárdenas, J., and Herrera, J.,** Inactivation and repression by ammonium of the nitrate reducing system in *Chlorella, Biochem. Biophys. Res. Commun.,* 38, 1009, 1970.
102. **Vennesland, B. and Jetschmann, C.,** The nitrate reductase of *Chlorella pyrenoidosa, Biochim. Biophys. Acta,* 227, 554, 1971.
103. **Herrera, J., Paneque, A., Maldonado, J. M., Barea, J. L., and Losada, M.,** Regulation by ammonia of nitrate reductase synthesis and activity in *Chlamydomonas reinhardii, Biochem. Biophys. Res. Commun.,* 48, 996, 1972.
104. **Lorimer, G. H., Gewitz, H.-S., Völker, W., Solomonson, L. P., and Vennesland, B.,** The presence of bound cyanide in the naturally inactivated form of nitrate reductase of *Chlorella vulgaris, J. Biol. Chem.,* 249, 6074, 1974.
105. **Aryan, A. P., Batt, R. G., and Wallace, W.,** Reversible inactivation of nitrate reductase by NADH and the occurrence of partially inactive enzyme in the wheat leaf, *Plant Physiol.,* 71, 582, 1983.
106. **Azuara, M. P. and Aparicio, P. J.,** *In vivo* blue-light activation of *Chlamydomonas reinhardii* nitrate reductase, *Plant Physiol.,* 71, 286, 1983.
107. **Subramanian, K. N. and Sorger, G. J.,** Regulation of nitrate reductase in *Neurospora crassa:* stability *in vivo, J. Bacteriol.,* 110, 538, 1972.
108. **Klein Amy, N. and Garrett, R. H.,** Repression of nitrate reductase activity and loss of antigenically detectable protein in *Neurospora crassa, J. Bacteriol.,* 144, 232, 1980.
109. **Moreno, C. G., Aparicio, P. J., Palacián, E., and Losada, M.,** Interconversion of the active and inactive forms of *Chlorella* nitrate reductase, *FEBS Lett.,* 26, 11, 1972.
110. **Palacián, E., de la Rosa, F., Castillo, F., and Gómez-Moreno, C.,** Nitrate Reductase from *Spinacea oleracea, Arch. Biochem. Biophys.,* 161, 441, 1974.
111. **Relimpio, A. M., Aparicio, P. J., Paneque, A., and Losada, M.,** Specific protection against inhibitors of the NADH-nitrate reductase complex from spinach, *FEBS Lett.,* 17, 226, 1971.
112. **Solomonson, L. P. and Vennesland, B.,** Properties of a nitrate reductase of *Chlorella, Biochim. Biophys. Acta,* 267, 544, 1972.
113. **Roldán, J. M. and Butler, W. L.,** Photoactivation of nitrate reductase from *Neurospora crassa, Photochem. Photobiol.,* 32, 375, 1980.
114. **Fritz, B. and Ninnemann, H.,** Photoreactivation by triplet flavin and photoinactivation by singlet oxygen of *Neurospora crassa* nitrate reductase, *Photochem. Photobiol.,* 41, 29, 1985.
115. **Vega, J. M., Herrera, J., Relimpio, A. M., and Aparicio, P. J.,** NADH-nitrate réductase de *Chlorella,* nouvelle contribution à l'étude de ses propriétés, *Physiol. Vég.,* 10, 637, 1972.

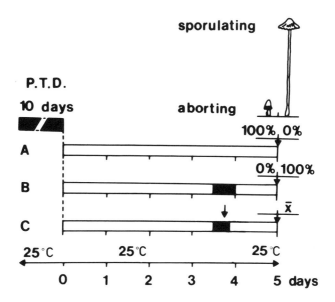

FIGURE 1. Main phases of fruiting body development of *Coprinus congregatus*. P.T.D., pretreatment in darkness. (A) The primordia were aborting in continuous light; (B) photoperiodic requirements for normal development of primordia as sporulating fruiting bodies; (C) effect of a blue-light break (arrow) on primordial development. x̄ Expresses the number of sporulating fruiting bodies per culture.

a complete agar medium and the second one with a liquid or agar-depleted medium. Primordial formation was induced by light when hyphae were growing on the depleted medium. Primordial initiation was strictly localized in the youngest hyphae of the culture and the primordia appeared on a ring just behind the front of growing hyphae at the time of illumination.[15] More than 300 primordia were photoinduced in a Petri dish and primordia initiation was assayed by counting the primordia under a binocular microscope 24 hr after light irradiation (Figure 3).

B. Photoinhibition of Primordial Development

A genuine photoperiodic response to light breaks was reported in *Coprinus*.[10] A blue light break during the inductive dark period inhibited the development of promordia, as did continuous light (Figure 1). As seen in Figure 2, the time of maximum sensitivity to light was at 3.2 hr after the beginning of darkness. On the other hand, a light break given one third of the way through the dark period or at the end of the dark period had no inhibitory effect on sporulation. The use of short light breaks (90 sec) of low fluence rate (38 mw m^{-2}) and the simultaneous irradiation of small cultures (Figure 4) offer the possibility of an action spectrum study for such a photoresponse.

III. ACTION SPECTRA

Action spectra for primordial photoinduction and photoinhibition of primordial development were determined using the Okazaki Large Spectrograph.[16] This spectrograph provides an intense photon fluence rate of monochromatic radiations between 250 and 800 nm.[17] The cultures were irradiated in threshold sample boxes. These boxes have seven compartments; each compartment received 40% of the fluence rate of the preceding one.[17] In this way, the determination of a fluence response curve required only a single irradiation period.

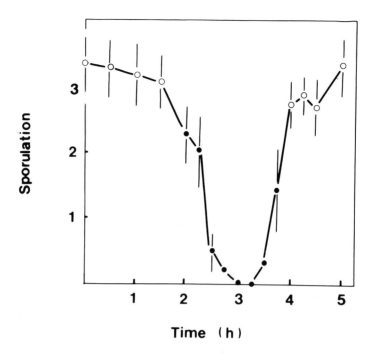

FIGURE 2.   Sporulating responses to blue light breaks (90 sec, 38 m · W · m⁻²) given at different times in a 5-hr dark period. The vertical lines indicate the 95% confidence interval. Closed symbols are used for means significantly different from the control (uninterrupted dark period) as shown by Student's *t* test at the 5% level and open symbols for means not significantly different from the control.

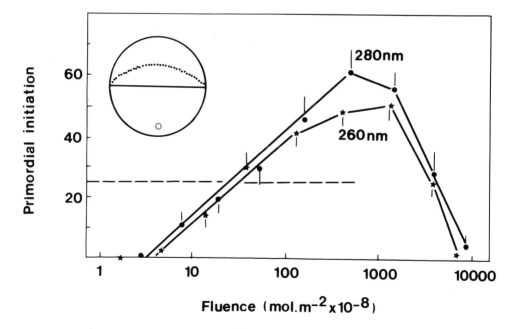

FIGURE 3.   Fluence response curves for primordial photoinduction in UV. The dotted line indicates the standard response. The drawing represents an experimental culture bearing photoinduced primordia symbolized by points.

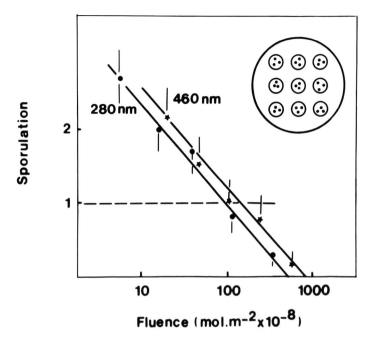

FIGURE 4.   Representative fluence response curves for photoinhibition of primordial development. The dotted line indicates the standard response. The drawing represents a Petri dish to be irradiated with nine small cultures bearing primordia symbolized by points.

A. Primordial Photoinduction

The photon fluence relationship for the photoinduction of primordia was studied at 26 wavelengths between 250 and 730 nm.[16] Representative curves calculated by the regression line method are given in Figure 3. A maximum number of primordia was induced by a photon fluence of 5 μmol m^{-2} (about 2 J m^{-2}) at 260 and 280 nm. Higher fluence levels reduced the number of induced primordia. These UV radiations (260 to 280 nm) had obviously two different effects on primordial initiation, i.e., an inductive one and an inhibitory one. According to the photon fluence, it was thus possible to distinguish between the well-known harmful effect of UV[18,19] and the photostimulatory effect of far UV. A saturation level of primordia was never obtained for near UV and visible radiations, even at the highest tested fluence rates.

The action spectrum was obtained by plotting the reciprocal of the number of incident photon fluences required to produce a 50% response (standard response) vs. wavelength. The action spectrum indicated peaks of activity at 260, 280, 370, and 440 nm. The relative quantum effectiveness at 280 nm was approximately four times higher than that of 440 nm. The upper limit of effectiveness for primordial photoinduction was 500 nm (Figure 5A).

B. Photoinhibition of Primordial Maturation

Fluence response curves for photoinhibition of primordial development were determined from 250 to 730 nm.[16] Agar blocks (18-mm diameter) bearing primordia and surrounding mycelium were removed from growing cultures and placed in darkness for 5 hr. A light break (60 sec) was given at the time of maximum sensitivity to light (Figure 2). The inhibitory effect of a light break on primordial development was measured by counting the number of sporulating fruiting bodies per culture. A maximum inhibitory effect (minimal sporulation) was observed when all the primordia aborted. The standard response was chosen as the

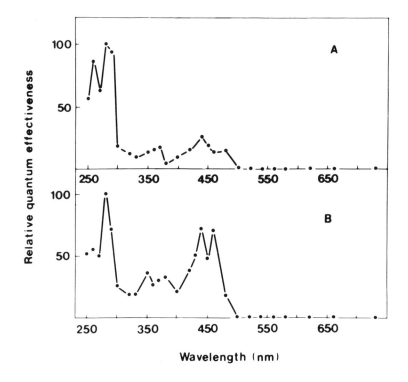

FIGURE 5.   Action spectra in *Coprinus*. (A) Primordial photoinduction; (B) photoinhibition of primordial development. The ordinate value for the most effective wavelength
was set at 100.

sporulation of one fruiting body per culture. This response was about 50% of the sporulation
in the uninterrupted dark control (Figure 4). The action spectrum for photoinhibition of
primordial development showed peaks of activity at 280, 350, 380, 440, and 460 nm. The
relative quantum effectiveness at 280 nm was about 1.3 times higher than that of blue light
(Figure 5B).

Slightly different peak positions were observed in the two action spectra (Figure 5). The
presence of screening pigments and the different localization of the photoreceptor in the two
responses could account for such differences. For primordial photoinduction, the white
growing hyphae are the photosensitive part of the colony.[15] For photoinhibition of primordial
development, the brown cap is the only photosensitive part of the primordia (unpublished
results).

Nevertheless, the general shape of the two action spectra was quite similar and suggested
a common blue-UV photoreceptor in the course of fruiting body morphogenesis in *C.
congregatus*. These action spectra showed typical features of the very common blue-UV
light responses.

C. The Photoreceptor

The nature of the blue-UV light photoreceptor (cryptochrome) has remained controver-
sial.[1,2,3,20] A review of a number of different action spectra for blue light responses suggested
that more than one blue light photoreceptor plays a role in morphogenesis of fungi.[3] On the
other hand, the shorter wavelength UV region has generally been scarcely investigated for
physiological effectiveness.[21,22]

The action spectra for phototropism in *Phycomyces*,[23,24] induction of perithecial formation
in *Gelasinospora reticulispora*,[25] and inhibition of conidial formation in *Alternaria solani*[26]
showed a sharp band of effectiveness at about 280 nm. Riboflavin shows an absorption

maximum in the far UV whereas β carotene does not.[27] An action spectrum band of great magnitude at 280 nm may thus be taken as additional evidence in support of a flavin photoreceptor and rules out the participation of carotene.[28] The strong peak of effectiveness at 280 nm for primordial photoinduction and photoinhibition of primordial development suggests the involvement of a flavoprotein photoreceptor in *Coprinus*.

IV. INHIBITOR STUDIES

Primordial photoinduction in *Coprinus* is a promising tool for studying biochemical changes following blue light perception. Primordial formation is a quickly occurring response taking place on a liquid medium in a predictable area of the colony.[15] The morphogenetical response involved changes in growth patterns. Initial events included the formation of hyphal lattices and the development of aerial hyphae and of complex mycelial aggregation.[29,30] Primordial formation was completed in less than 24 hr.[15]

The effects of various flavin inhibitors and of inhibitors of nucleic acid and protein synthesis were studied on the light-induced primordial formation of *Coprinus*.

A. Flavin Inhibitors

Many of the photochemical reactions of riboflavin occur through the long-lived triplet state.[27] Potassium iodide and azide are effective in depopulating the excited state of flavins. An effect of these drugs on blue light physiological responses can be taken as additional evidence for a flavin photoreceptor. Potassium iodide was found to be a specific inhibitor of the inverse phobic response of *Euglena* and of phototropism in corn seedlings.[2] Moreover, phenylacetic acid, a molecule known to bind covalently to irradiated flavins inhibited phototropism in corn seedlings.[31] However, the use of chemical inhibitors has many drawbacks as long as their specificity is not demonstrated. Experiments were carried out in *Coprinus* to ascertain the specificity of flavin inhibitors on a blue light response (primordial initiation), using sclerotial formation as a control. Sclerotial formation took place in darkness; a great number of sclerotia differentiated in youngest hyphae of the culture when the medium was supplemented with $(NH_4)_2 HPO_4$.[15] Genetic studies suggested that primordial and sclerotial formation in *Coprinus* shared a common pathway of differentiation.[32]

The results of Table 1 clearly show that potassium iodide did not inhibit blue photoinduced primordial formation. Phenylacetic acid and sodium azide inhibited the primordial and sclerotial formation. The activity of phenylacetic acid on the nonphotoinduced process of sclerotial formation in *Coprinus* leads to the conclusion that this drug does not act at the photoreceptor level. Schönbohm and Schönbohm[23] demonstrated that, in *Mougeotia*, the effects of potassium iodide were not achieved by quenching flavins but rather via inhibiting secondary processes. Such experiments do not rule out the participation of a flavin photoreceptor; they demonstrated the uselessness of these flavin inhibitors in photoreceptor identification.

B. Inhibitors of Nucleic Acid and Protein Synthesis

The relationships between photoinduction and RNA and protein synthesis have been studied in photoinduced primordial initiation of *Coprinus*. Indirect implication of nucleic acid and protein involvement came from studies using the inhibitors 5-fluorouracil and cycloheximide.[15] A short-time application of these inhibitors at $5.10^{-5}M$ completely suppressed light-induced primordial formation and greatly reduced the chemically induced sclerotial formation (Table 1).

Experiments on the time of application of inhibitors showed that fluorouracil inhibited to a greater extent primordial formation when added during the photoinduction period, whereas cycloheximide inhibited primordial formation only when added after the photoinduction

Table 1
EFFECT OF VARIOUS INHIBITORS ON PRIMORDIAL AND SCLEROTIAL FORMATION IN *COPRINUS CONGREGATUS*

Inhibitor (M)		Primordial production	Sclerotial production
Phenyl acetic acid	10^{-2}	26	—
	5.10^{-3}	50	25
	10^{-3}	82	52
	5.10^{-4}	82	88
NaN_3	10^{-3}	40	0
	10^{-4}	75	38
	10^{-5}	76	70
	10^{-6}	105	101
KI	10^{-2}	102	—
	5.10^{-3}	107	—
	10^{-3}	100	—
5-Fluorouracil	5.10^{-5}	0	1
	10^{-5}	26	35
	5.10^{-6}	80	59
	10^{-6}	92	78
	5.10^{-7}	110	85
Cycloheximide	5.10^{-5}	0	6
	$2.5.10^{-5}$	15	—
	10^{-5}	103	24
	5.10^{-6}	—	54
	10^{-6}	110	94

Note: Inhibitors were applied for 3.5 hr during primordial photoinduction (3 hr of blue light) and for 6 hr during chemically induced slerotial formation ((NH_4)$_2$ HPO_4 7.6 mM for 6 hr). Primordial and sclerotial productions are expressed as percentage of the controls without inhibitors.

period (Table 2). These results presented evidence for the requirement of *de novo* RNA synthesis followed by protein synthesis in primordial photoinduction. Such a sequential synthesis was suggested to occur for different photoresponses in fungi such as conidiation in *Trichoderma* and *Botrytis*.[4,34]

However, such indirect evidence using inhibitors is not a conclusive evidence. Evidence is needed for distinct changes in transcriptional levels and distinct translational products. In *Coprinus*, different specific proteins were synthesized during light-induced primordial formation and during chemically induced sclerotial formation (unpublished results). Preliminary results using ^3H leucine showed a specific incorporation of labeling in protein synthesized after light induction (unpublished results). A direct approach in finding new and different mRNAs during photoinduced sporulation of *Trichoderma* was not successful.[35] The results did not preclude transcriptional changes but indicated that the differences may be quantitatively small. The extensive work carried out in *Schizophyllum commune* to demonstrate differences in proteins and RNAs at the time of basidiocarp formation was not primarily focused on photodifferentiation.[36,37] Nevertheless, during the formation of fruiting body primordia, a small number of novel abundant mRNAs appeared in the RNA population; about 5% of the RNA mass present in the fruiting dikaryon were absent in the vegetative monokaryon.[36]

Table 2
EFFECT OF TIME OF APPLICATION OF INHIBITORS ON PHOTOINDUCED PRIMORDIAL FORMATION OF *COPRINUS CONGREGATUS*

Inhibitor	Before photoinduction	During photoinduction	After photoinduction
5-Fluorouracil (10^{-5} M)	77.9	26	85.4
Cycloheximide (10^{-5} M)	103.6	102.7	44.7

Note: Primordial formation is expressed as percentage of the controls without inhibitors. The duration of inhibitor application was 3.5 hr and the length of blue-light induction period 3 hr.

V. CONCLUSIONS

The research on the nature of cryptochrome has long been focused on the controversy between flavins and carotenes. Many arguments have been accumulated to favor the flavin hypothesis.[1,38] The strong peak of effectiveness at 280 nm in action spectra for primordial photoinduction and photoinhibition of primordial development suggests the involvement of a flavoprotein as a blue-UV photoreceptor in *Coprinus*.

Experiments using inhibitors known to react with illuminated flavins demonstrated that these inhibitors do not act at the photoreceptor level and can no longer be regarded as specific inhibitors of blue-UV light responses. The most promising approach to elucidate the nature of cryptochrome is to use a system in which the photoreceptor level can be perturbated. Riboflavin-requiring mutants have been isolated in *Phycomyces*,[6] *Neurospora*,[39] and *Trichoderma*.[40] Experiments are carried out to isolate such mutants in *Coprinus*. Biochemical and spectroscopic analysis of such strains should lead to elucidation of cryptochrome and of the primary events of the phototransduction chain.[35]

Most of the studies on photodifferentiation in fungi have used whole mycelia with different parts not uniformly competant to be induced by light. On the other hand, qualitative changes in transcription of mRNAs and translation to proteins in photoinduced mycelia may be quantitatively small and difficult to detect.[4,36] The new cultural procedure for primordial photoinduction of *Coprinus* overcomes this problem. Therefore, this morphogenetic response will be a promising tool for studying molecular events occurring during photomorphogenesis.

REFERENCES

1. **Senger, H., Ed.,** *The Blue Light Syndrome,* Springer Verlag, Berlin, 1980.
2. **Schmidt, W.,** Physiological blue light reception, in *Structure and Bonding,* Vol. 41, Dunitz, J. D., Goodenough, J. B., Hemmerich, P., Ibers, J. A., Jorgensen, C. K., Neilands, J. B., Reinen, D., and Williams, R. J. P., Eds., Springer-Verlag, Berlin, 1980, 1.
3. **Briggs, W. R. and Iino, M.,** Blue light absorbing photoreceptors in plants, *Phil. Trans. R. Soc. London B,* 303, 347, 1983.
4. **Gressel, J. and Rau, W.,** Photocontrol of fungal development, in *Encyclopedia of Plant Physiology,* Vol. 16B, *Photomorphogenesis,* Pirson, A. and Zimmermann, H., Eds., Springer-Verlag, Berlin, 1983, 603.
5. **Rau, W.,** Blue light-induced carotenoid biosynthesis in microorganisms, in the *Blue Light Syndrome,* Senger, H., Ed., Springer-Verlag, Berlin, 1980, 283.
6. **Galland, P. and Lipson, E. D.,** Photophysiology of *Phycomyces blakesleanus, Photochem. Photobiol.,* 40, 795, 1984.

7. **Russo, V. A.,** Sensory transduction in phototropism: genetic and physiological analysis in *Phycomyces,* in *Photoreception and Sensory Transduction in Aneural Organisms,* Lenci, F. and Colombetti, G., Eds., Plenum Press, New York, 1980, 373.

8. **Degli-Innocenti, F. and Russo, V. E. A.,** Isolation of new white collar mutants of *Neurospora crassa* and studies on their behavior in the blue light-induced formation of protoperithecia, *J. Bacteriol.,* 159, 757, 1984.

9. **Durand, R. and Jacques, R.,** Action spectra for fruiting of the mushroom *Coprinus congregatus, Arch. Microbiol.,* 132, 131, 1982.

10. **Durand, R.,** Photoperiodic response of *Coprinus congregatus:* effects of light breaks on fruiting, *Physiol. Plant.,* 55, 226, 1982.

11. **Durand, R.,** Light breaks and fruit-body maturation in *Coprinus congregatus:* dark inhibitory and dark recovery process, *Plant Cell Physiol.,* 24, 899, 1983.

12. **Manachère, G.,** Recherches physiologiques sur la fructification de *Coprinus congregatus* Bull. ex Fr. Action de la lumière. Rythme de production de carpophores, *Ann. Sc. Nat. Bot.,* 11, 1, 1970.

13. **Robert, J. C. and Durand, R.,** Light and temperature requirements during fruit-body development of a Basidiomycete mushroom, *Coprinus congregatus, Physiol. Plant.,* 46, 174, 1979.

14. **Kamada, T., Kurita, R., and Takemaru, T.,** Effects of light on basidiocarp maturation in *Coprinus macrorhizus, Plant Cell Physiol.,* 19, 263, 1978.

15. **Durand, R.,** Effects of inhibitors of nucleic acid and protein synthesis on light-induced primordial initiation in *Coprinus congregatus, Trans. Br. Mycol. Soc.,* 81, 553, 1983.

16. **Durand, R. and Furuya, M.,** Action spectra for stimulatory and inhibitory effects of UV and blue light on fruit-body formation in *Coprinus congregatus, Plant Cell Physiol.,* 26, 1775, 1985.

17. **Watanabe, M., Furuya, M., Miyoshi, Y., Inoue, Y., Iwahashi, I., and Matsumoto, K.,** Design and performance of the Okazaki Large Spectrograph for photobiological research, *Photochem. Photobiol.,* 36, 491, 1982.

18. **Leach, C. M.,** A practical guide to the effects of visible and ultraviolet radiation on fungi, in *Methods in Microbiology,* Vol. 4, Booth, C., Ed., Academic Press, New York, 1971, 609.

19. **Thomas, G.,** Effects of near ultraviolet light on microorganisms, *Photochem. Photobiol.,* 26, 669, 1977.

20. **Senger, H.,** The effect of blue light on plants and microorganisms, *Photochem. Photobiol.,* 35, 911, 1982.

21. **Wellmann, E.,** UV radiation in photomorphogenesis, in *Encyclopedia of Plant Physiology,* Vol. 16B, Shropshire, W. and Mohr, H., Eds., Springer-Verlag, Berlin, 1983, 745.

22. **Inoue, Y.,** Re-examination of action spectroscopy in blue, near-UV light effects, in *Blue Light Effects in Biological Systems,* Senger, H., Ed., Springer-Verlag, Berlin, 1984, 110.

23. **Curry, G. M. and Gruen, H. E.,** Action spectra for the positive and negative phototropism of *Phycomyces* sporangiophores, *Proc. Natl. Acad. Sci. U.S.A.,* 45, 797, 1959.

24. **Delbrück, M. and Shropshire, W., Jr.,** Action and transmission spectra of *Phycomyces, Plant Physiol.,* 35, 194, 1960.

25. **Inoue, Y. and Watanabe, M.,** Perithecial formation in *Gelasinospora reticulispora.* VII. Action spectra in UV region for the photoinduction and the photoinhibition of photoinductive effect brought by blue light, *Plant Cell Physiol.,* 25, 107, 1984.

26. **Honda, Y. and Nemoto, M.,** An action spectrum for photoinhibition of conidium formation in the fungus *Alternaria solani, Can. J. Bot.,* 62, 2865, 1984.

27. **Presti, D. and Delbrück, M.,** Photoreceptors for biosynthesis, energy storage and vision, *Plant Cell Environ.,* 1, 81, 1978.

28. **De Fabo, E. C., Harding, R. W., and Shropshire, W., Jr.,** Action spectrum between 260 and 800 nm for the photoinduction of carotenoid biosynthesis in *Neurospora crassa, Plant Physiol.,* 57, 440, 1976.

29. **Matthews, T. R. and Niederpruem, D. J.,** Differentiation in *Coprinus lagopus.* I. Control of fruiting and cytology of initial events, *Arch. Microbiol.,* 87, 257, 1972.

30. **Henderson, H. E. and Ross, I. K.,** Ultrastructural studies of vegetative and fruiting mycelia of *Coprinus congregatus, Mycologia,* 75, 634, 1983.

31. **Schmidt, W., Hart, J., Filner, P., and Poff, K. L.,** Specific inhibition of phototropism in corn seedlings, *Plant Physiol.,* 60, 736, 1977.

32. **Moore, D.,** Developmental genetics of *Coprinus cinereus:* genetic evidence that carpophores and sclerotia share a common pathway of initiation, *Curr. Genet.,* 3, 145, 1981.

33. **Schönbohm, E. and Schönbohm, E.** Multiple effects of the flavin quencher potassium iodide on light-dark-processes in the green alga *Mougeotia,* in *Blue Light Effects in Biological Systems,* Senger, H., Ed., Springer-Verlag, Berlin, 1984, 137.

34. **Tan, K. K.,** Light-induced fungal development, in *The Filamentous Fungi,* Vol. 3, Smith, J. E. and Berry, D. R., Eds., Arnold, London, 1978, 334.

35. **Gressel, J.,** The quest for *Trichoderma* cryptochrome, in *Blue Light Effects in Biological Systems,* Senger, H., Ed., Springer-Verlag, Berlin, 1984.

36. **Hoge, J. H. C., Springer, J., and Wessels, J. G. H.,** Changes in complex RNA during fruit-body initiation in the fungus *Schizophyllum commune, Exp. Mycol.,* 6, 233, 1982.
37. **De Vries, O. M. H. and Wessels, J. G. H.,** Patterns of polypeptide synthesis in non-fruiting monokaryons and a fruiting dikaryon of *Schizophyllum commune, J. Gen. Microbiol.,* 130, 145, 1984.
38. **Senger, H.,** *Blue Light Effects in Biological Systems,* Springer-Verlag, Berlin, 1984.
39. **Paietta, J. and Sargent, M. L.,** Photoreception in *Neurospora crassa:* correlation of reduced light sensitivity with flavin deficiency, *Proc. Natl. Acad. Sci. U.S.A.,* 78, 5573, 1981.
40. **Horwitz, B. A. and Gressel, J.,** Elevated riboflavin requirement for postphotoinductive events in sporulation of a *Trichoderma* auxotroph, *Plant Physiol.,* 71, 200, 1983.

Chapter 5

BLUE LIGHT CONTROL OF PIGMENT BIOSYNTHESIS

W. Rau and E. L. Schrott

TABLE OF CONTENTS

I. INTRODUCTION

Light is one of the most important environmental factors for all organisms. In chlorophyll-containing plants it serves as the source of energy for the light-harvesting photosynthesis. However, irradiation, particularly with ultraviolet (UV) and blue light, may also cause deleterious effects in organisms. Since plants and microorganisms are unable to escape harmful situations by migration — at least not by long-distance migration — it is not astonishing that these organisms have developed mechanisms for the photoregulation of both development and metabolism. Among these adaptations, photoregulation of biosyntheses is an ubiquitously observed phenomenon. The most striking example of a drastic effect of light on a biosynthetic pathway is the obligatory light dependence of chlorophyll formation in angiosperms (see Chapter 7).

Among the mass pigments occurring in the plant kingdom, carotenoids are obviously the most widespread ones; not only do all photosynthetic organisms contain carotenoids, but also the coloring of many fungi and bacteria are caused by these pigments. Carotenoids, although playing a less obvious role in plants than chlorophyll, have many distinct functions in several groups of organisms; the most important ones should be mentioned here briefly. In photosynthesis several possible functions of carotenoids have been discussed; their role as accessory pigments in some groups of algae is well documented and they appear to be constituents of the photosystems. There is good evidence that carotenoids protect cells of green plants, fungi, and bacteria against the potentially harmful effects of irradiation. Carotenoid pigments are responsible for many of the yellow and red colors of flowers and fruits and may help to attract animals needed for the distribution of both pollen and seeds. In fungi they play a role in sexual reproduction. The reader interested in details of these and additional functions is referred to pertinent reviews.[1-3]

In contrast to animals, plants and bacteria have the capacity for *de novo* biosynthesis of carotenoids. Although results and problems of the biosynthesis of carotenoids have been reviewed extensively in the past years,[4-6] for the convenience of the reader and for a better understanding of the following sections of this chapter, the main steps of the biosynthetic pathway are summarized here. From the first specific precursor isopentenyl-pyrophosphate, a C_5-compound, the C_{20}-compound geranylgeranyl-pyrophosphate is formed by step-wise addition. Tail-to-tail addition of two of these molecules leads to the first compound with the typical C_{40}-skeleton of carotenoids, phytoene. Since the molecule carries only three conjugated double bonds, phytoene absorbs UV light only and therefore is not colored. By way of step-wise desaturation (increasing the number of conjugated double bonds), ring closure, and finally by introduction of other groups (in the case of xanthophylls different oxygen functions), the large variety of carotenoids is formed. The first steps of this pathway follow the Porter-Lincoln scheme.[7] As an example for this scheme, the carotenoids formed by some fungi are shown in Figure 1. In these fungi, biosynthesis is blue light dependent.

Considering the functions as well as the widespread occurrence of carotenoids, it is not surprising that biosynthesis of carotenoid pigments is under photocontrol in many organisms. Blue light control of carotenoid biosynthesis has so far been reported for algae, fungi, and bacteria. However, the extent of synthesis regulated by light differs from organism to organism. Accepting some simplification we may distinguish three types. In many organisms, appreciable amounts of carotenoids are synthesized in complete darkness and illumination only enhances the rate of synthesis. This appears to be the case in phytochrome-mediated carotenoid accumulation in higher plants, but no blue light effect is of this type. In other species, illumination is obligatory for a substantial synthesis of certain carotenoids but accumulation of these pigments proceeds only during illumination; that is, the permanent presence of the inducing factor "light" is essential for the response. Photoregulation of carotenogenesis in the fungus *Fusarium aquaeductuum* (Figure 2) may serve as an example

FIGURE 1. Carotenoids synthesized by some fungi arranged according to the Porter-Lincoln scheme.

for the third type. The accumulation rate is very low in the dark; already a brief exposure to light — i.e., a few minutes — induces substantial pigment production in the subsequent dark period only for a certain period of time. This type of photocontrol exhibits all features of a "classical" induction mechanism; that is, light serves only as a "trigger" and all dark reactions are consistently a strict consequence of photoinduction. The photoregulation of carotenoid biosynthesis poses several questions, e.g., which are the photoreceptors responsible for this reaction, which step(s) of the biosynthetic pathway are under photocontrol, and what mechanism(s) are involved in the transformation of the external signal — light — into the physiological and biochemical response, i.e., synthesis of pigments. Though in the past decades a great deal of work on these problems has been accomplished, at the moment we still are far from a complete insight. It is not the intention of this chapter to give a comprehensive survey of the literature relevant to this topic but rather to summarize the main characteristics of the reaction, in particular that of the mechanism of photoregulation. The reader interested in more detail is referred to previous reviews on the photocontrol of carotenoid biosynthesis with regard to various aspects.[8-12]

II. THE PHENOMENA

A. Higher Plants

Light stimulation of carotenoid biosynthesis in higher plants has been reported so far only for angiosperms. In young leaves carotenoids are synthesized exclusively in plastids as integral constituents of the plastid membranes. Etiolated seedlings contain some carotenoids in the plastids, the majority of which are xanthophylls. It is a well-documented phenomenon,

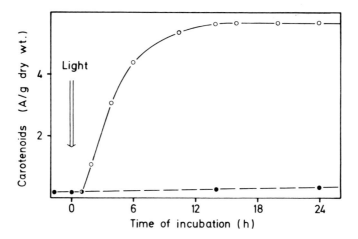

FIGURE 2.   Accumulation of carotenoids in young mycelia of *Fusarium aquaeductuum* in the dark (closed circles) and following a brief illumination (open circles). The arrow indicates the period of illumination. (From Rau, W., *Pure Appl. Chem.*, 57, 782, 1985.)

known for many years and investigated by numerous authors, that light-grown plants contain higher amounts of carotenoids compared to dark-grown plants (see Reference 13).

In ripening tomato and paprika fruits, photoregulation of the transformation of chloroplasts into chromoplasts, accompanied by a massive synthesis of carotenoids was found.[14,15] The experimental data so far available indicate in all these cases an induction mechanism by way of phytochrome. In addition to phytochrome, blue-light effects seem to be essential for the production of the carotenoids of mature chloroplasts. For example, in wheat leaves induction of carotenogenesis may somehow depend on light absorption by protochlorophyll(ide);[16] also in mustard seedlings the role of chlorophyll in carotenoid synthesis has been elucidated[17] (see also Section III.A.2).

Numerous phytochrome-type responses in higher plants interact with blue light responses[18,19] (also see this volume). Therefore, it seems to only be a matter of time before such interactions are also found for carotenoid biosynthesis in some higher plants, since evolution appears to have made use of all potent photoreceptor systems available.

B. Algae

In marked contrast to the uniformity of carotenoid patterns in green tissues of higher plants, the chloroplast carotenoids of the different classes of algae show considerable qualitative differences. Therefore, they serve as excellent chemotaxonomic characters in the classification of algae (see References 20 and 21).

In algae, photoregulation of carotenoid biosynthesis is very rare. The only case in wild-type strains seems to be that of the phytoflagellate *Euglena gracilis*. Like angiosperms, it requires light for both chlorophyll synthesis and development of a functioning chloroplast. Although dark-grown cultures are able to synthesize all carotenoids present in light-grown organisms, the amount of carotenoids is greatly reduced when the algae are cultured heterotrophically in the dark. Organisms "bleached" by mutation or by inhibitors lack the xanthophylls found in the wild type and show photostimulation of carotenogenesis.[22-26] Therefore, the situation appears to be similar to that in higher plants. This is also supported by data obtained from experiments with the herbicide SAN 9789, a compound which inhibits the synthesis of normal carotenoids and causes the accumulation of phytoene.[27] Evidence has been presented that in *Euglena* the synthesis of carotenoids and chlorophyll are strongly correlated. Moreover, electron microscopic studies indicated that inhibition of carotenoid production prevented thylakoid membrane assembly.

The unicellular green alga *Chlorella* produces intact chloroplasts and the usual plastid carotenoids when grown in the dark. However, there are some early reports indicating that illumination causes quantitative and qualitative differences in carotenoid contents,[28] but strong photoregulation has only been detected in and investigated in more detail with *Chlorella* mutants. In a chlorophyll-free yellow mutant, carotenoid biosynthesis is enhanced by blue light.[29] A green mutant accumulates a series of acyclic polyenes in the dark; cyclic carotenes are synthesized only during illumination.[30,31]

More recently photoregulatory mutants from another species of green algae, *Scenedesmus obliquus,* have been examined. A chemically induced mutant strain synthesized appreciable amounts of chlorophyll and cyclic carotenoids only when illuminated. In dark-grown cultures mainly ζ-carotene was accumulated; during a subsequent period of the illumination, cyclic carotenes and xanthophylls were formed at the expense of the accumulated precursors.[32-35] The authors concluded from their results that as in *Euglena,* the presence of xanthophylls is important for the development of normal chloroplasts structures.

Again in connection with the light-dependent development of the photosynthetic apparatus, the production of carotenoids was examined during greening of an X-ray-induced mutant.[36,37] Grown in darkness, the mutant contained only acyclic carotenes in addition to trace amounts of chlorophyll a; during a 24-hr period of greening in white light, both cyclic carotenes and xanthophylls were formed. Activity of photosystem II of photosynthesis was only found after xanthophylls had been produced, as also demonstrated by inhibitor studies using nicotine and anaerobiosis. The authors, therefore, concluded that xanthophylls are needed for structural organization of photosystem II. In addition, cyclic carotenoids proved to be essential for photoprotection of the chlorophylls.

C. Fungi and Bacteria

Photoregulation of carotenoid biosynthesis has been reported for a number of fungi. In some species, such as *Phycomyces*, light increases the rate of pigment production, whereas in others, illumination is obligatory for distinct coloring. The species of fungi and bacteria so far known to show photoregulation of carotenogenesis have been compiled previously.[8,38,39]

Many of the results obtained from investigations using these organisms have contributed in particular a great deal to the elucidation of the mechanism of photoregulation. These data will be presented in the following chapters, but the main features of the phenomenon are summarized here briefly.

Those fungi and bacteria that obligatorily require light for a massive production of carotenoids synthesize only trace amounts of pigments when grown in the dark. Even a brief exposure to light (i.e., a few seconds or minutes) induces substantial carotenogenesis, although higher doses of illumination are necessary for bulk production. Light saturation of the response has been found for *Mycobacterium* sp., *Mycobacterium marinum*, and, in earlier reports, for *Neurospora crassa*, but no saturation could be achieved for *Fusarium aquaeductuum* (see Reference 10). A recent investigation of the dose-response relationship in *N. crassa* revealed that prolonged illumination after a first saturation led to a second photoinduction which seemed not to be saturable.[40]

In all species studied so far, the time courses of photoinduced pigment accumulation are very similar to that shown in Figure 2 for *Fusarium,* but the absolute length of the distinct periods differ individually. A short illumination is followed by a lag period before the amount of carotenoids increases rapidly for a certain period of time. Thereafter pigment production ceases. Under continuous illumination carotenogenesis does not stop but instead continues with a reduced but linear rate.

Summing up, results gathered from experiments with brief illumination periods indicate that signal transduction and subsequent dark reactions consistently are a consequence of photoinduction. Thus photoregulation in these organisms exhibits all features of a classical induction mechanism as mentioned in the introduction to this chapter. Therefore, these organisms are a useful model system to study the sequence of events starting with the absorption of light and ending up with the synthesis of carotenoids, termed the "mechanism" of the photoregulation in the following. From the results described as well as from results of studies using inhibitors of protein synthesis it has been concluded that photoinduction leads to a *de novo* synthesis of the carotenogenic enzymes. Evidence for these conclusions will be presented in the following sections of this chapter.

III. MECHANISMS OF CONTROL

A. Photoreceptors and Signal Transduction

The usual approach for elucidation of the photoreceptor responsible for a certain response is a comparison of the action spectrum of the response with the absorption spectra of the putative pigment. However, such a comparison has to consider factors which might influence the shape of the action spectrum, e.g., screening by different pigments or fluorescence. It must also be considered that the absorption spectrum of the photoreceptor might be changed *in situ*. Since these obstacles will be discussed in detail in the chapters on primary photoreceptors, action spectra, and spectroscopy, possible photoreceptors for carotenoid biosynthesis will be presented here without detailed discussions. Instead, some emphasis will be given to results which might help to elucidate the signal transduction, that is, the problem of how the organism can translate the physical signal "light" into a biochemical signal.

1. Cryptochrome

Besides earlier reports on the spectral dependence of carotenoid biosynthesis in fungi, Zalokar[41] was the first to determine an action spectrum of carotenogenesis in the spectral region between 400 and 500 nm for nonconidiating cultures of *Neurospora crassa;* prevention of conidiation is important because in conidia carotenoid production is not light dependent. Later, action spectra including shorter wavelengths were elaborated for this fungus as well as for *Fusarium aquaeductuum* and *Mycobacterium* sp. as illustrated by Figure 3. These exhibit some common characteristics: a maximum at 370 to 380 nm and three peaks, or at least shoulders, between 400 and 500 nm, and the fact that light with wavelengths longer than approximately 520 nm proved to be ineffective.

The shapes of these action spectra of carotenoid biosynthesis resemble those obtained for a variety of developmental and movement responses in a large number of different plant species among which the phototropic reaction in coleoptiles and sporangiophores are the most prominent and which are now named cryptochromal. Arguments in favor of each of the two candidates to be the photoreceptor responsible — flavins or carotenoids — are discussed in detail in other chapters of this book.

In fungi and nonphotosynthetic bacteria, the photoreaction has been found to be independent of temperature, indicating that the primary event is a photochemical reaction (see References 10 and 11). A brief exposure to light is already sufficient to induce carotenoid synthesis, although higher fluences are required for bulk pigment production. For a brief exposure the Bunsen-Roscoe law of reciprocity has proved to be valid (see References 11 and 40). Using prolonged periods of illumination, the amount of pigment produced is proportional to the logarithm of the quantity of incident light in *F. aquaeductuum*,[42] whereas in *N. crassa*,[40] as in *Phycomyces blakesleeanus*,[45] biphasic fluence-response curves have been observed; in the latter case, the involvement of two photoreceptors was therefore suggested.

At present, signal transduction resulting from light absorption by the photoreceptor is far from being completely understood. Since in some investigations carotenogenesis is used as a model system for the elucidation of this problem, the major results are described here without quoting all of the details.

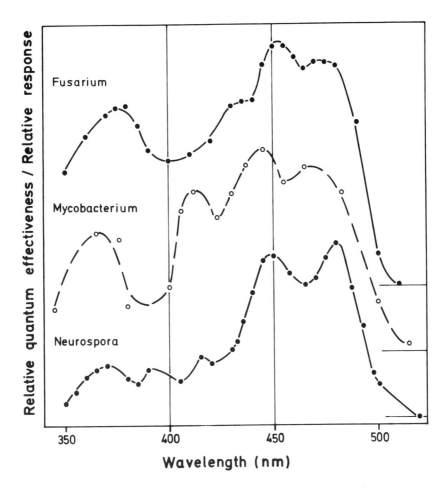

FIGURE 3. "Cryptochromal"-type action spectra of photoinduced carotenoid biosynthesis in *Fusarium aquaeductuum* (redrawn from Rau, W.[42]), *Mycobacterium sp.* (redrawn from Howes, C. D. and Batra, P. P.,[43]) and *Neurospora crassa* (redrawn from De Fabo, E., et al.[44]). The base lines have been shifted as indicated.

The presence of oxygen is essential for optimum photoinduction, although minor induction in the absence of O_2 has been found in some fungi but not in *Mycobacterium* (see Reference 11). In *Fusarium,* saturation of the photoreaction under anaerobic conditions is reached at a relatively low fluence and is independent of both fluence rate and time of illumination. Moreover, when mycelia were illuminated in a nitrogen atmosphere with saturating fluences, subsequently supplied with oxygen in the dark, and then illuminated a second time on the absence of oxygen, they were susceptible to an additional photoinduction. Obviously, the photoreceptor system can be reactivated by oxygen in the dark. We therefore concluded that oxygen functions as an electron acceptor keeping the photoreceptor in a proper state of oxidation.[46] In contrast, other investigators (see Reference 8) have proposed that oxygen participates directly in the primary photoprocess. As in photodynamic effects, the involvement of singlet oxygen might be taken into consideration.

A strong reducing substance, dithionite, applied to mycelia prior to or shortly after illumination inhibits photoinduced carotenoid formation completely and specifically (Figure 4); the effectiveness of dithionite decreases gradually with the time elapsed between illumination and the application of the substance and is completely lacking 20 min after the onset illumination. On the other hand, incubation of the mycelia with buffered hydrogen peroxide

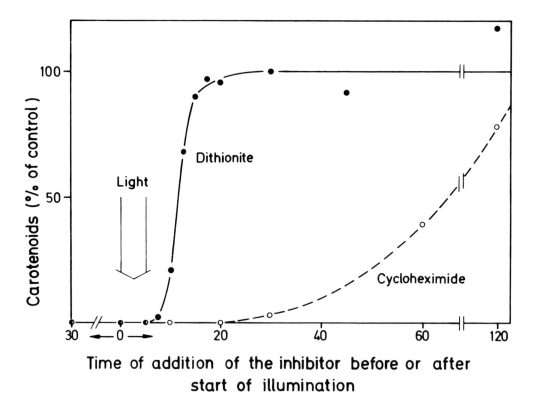

FIGURE 4.   Comparison of the inhibitory effects on photoinduced carotenogenesis of dithionite (5.10^{-3} *M*) and of cycloheximide ($6{\cdot}10^{-5}$ *M*) applied to mycelia of *Fusarium aquaeductuum* at various times before or after illumination. Dithionite was removed 30 min after addition by rinsing the mycelia with buffer. (Redrawn from Theimer, R. R. and Rau, W., *Planta* 92, 129, 1970; Rau, W. and Rau-Hund, A., *Planta*, 136, 49, 1977; and from unpublished results.)

solution in the dark causes induction of carotenogenesis, which indicates that hydrogen peroxide may, at least to some extent, substitute for light.[47]

Comparison with the time scale of the decreasing effect of cycloheximide (see Section III.B.2.d) under the same experimental conditions (Figure 4) makes it quite clear that dithionite affects a very early step in the reaction chain.

As mentioned above, red light is not effective to induce carotenogenesis in *Fusarium*. However, when the mycelia are incubated with methylene blue, toluidine blue, or neutral red (dyes which absorb red light), and are simultaneously illuminated with red light, carotenoid biosynthesis is induced. This photoinduction by red light is triggered only in the presence of these photodynamically active redox dyes, whereas nonredox dyes such as dichlorophenolindophenol or malachite green are not effective.[49] These results led us to suggest that the dyes may act as artificial photoreceptors. However, the question is still open whether they are bypassing the natural photoreceptor or whether they act as additional receptors transferring the energy to the natural photoreceptor; results obtained by Britz et al.[50] tend to support the former possibility.

Further evidence for a primary photoact proceeding the subsequent biochemical events comes from results reveiling a "memory" phenomenon. When organisms such as *Fusarium*, *Neurospora*, and others (see Reference 11) are illuminated at a low temperature, i.e., near 0°C, no carotenoids are formed during the following dark period when they are left in the cold. However, when transferred to room temperature after some hours, they begin to synthesize the pigments induced. This "memory" indicates that the photoact leads to a stable "inductive state" and may be a powerful instrument for further investigations.

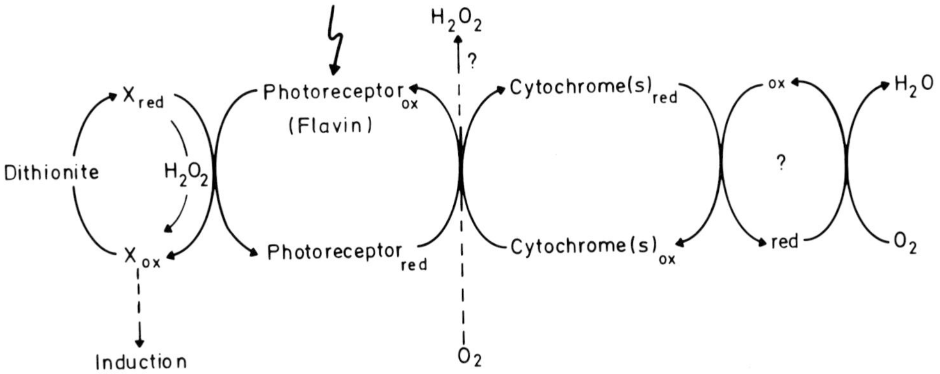

FIGURE 5. Scheme illustrating the proposed events during signal transduction mediated by cryptochrome. (From Rau, W., in *The Blue Light Syndrome,* Senger, H., Ed., Springer-Verlag, Berlin, 1980, 283. With permission.)

In various plants and microorganisms, illumination with blue light both in vivo and in vitro induces absorbance changes which were attributed to the photoreduction of a b-type cytochrome via a flavin moiety. The relevance of these light-induced absorbance changes (LIACs) for blue light perception and signal transduction has been discussed in recent reviews and will be dealt with in another chapter of this book.

Summarizing the data for photoinduced carotenogenesis and also considering the results for other phenomena mediated by blue light, we have drawn a hypothetical scheme illustrating the events during photoinduction via cryptochrome and the subsequent signal transduction[10] (Figure 5). In this scheme it is assumed that photoreduction of a flavin photoreceptor causes a concomitant oxidation of a yet hypothetical compound "X_{red}"; the resulting "X_{ox}" is stabilized rapidly by subsequent reactions yielding a "photooxidation" product. The reduced flavin photoreceptor may be reoxidized by transferring electrons either to a cytochrome (LIACs) and finally to oxygen or directly to oxygen. For other cryptochrome-mediated responses similar schemes have been discussed.[51] For the signal transduction steps towards biosynthesis of carotenoids, two "key substances" may be considered as the triggers: reduced components (cytochrome?) or the oxidized component "X_{ox}". Taking into account the data on the effect of dithionite and H_2O we consider a stabilized "X_{ox}" to be the more likely candidate.

2. Chlorophyll

It has been presumed that in seedlings of higher plants light absorbed by photoreceptors other than phytochrome is necessary for the formation of typical carotenoids of chloroplasts. Action spectra for the induction of carotenoid synthesis were determined in wheat leaves[16] and in the unicellular algae;[52] from these data the authors concluded that carotenogenesis may depend on the light absorption by protochlorophyll(ide). Frosch and Mohr[17] have shown that in seedlings of white mustard carotenogenesis was stimulated via phytochrome, and in addition carotenoid accumulation was further increased when large amounts of chlorophyll were synthesized. They explained this effect by the concept that in the plastids grana formation can only proceed when chlorophyll is available and, as a consequence, large amounts of carotenoids are incorporated into the thylakoid membranes. This then decreases a "free carotenoid pool", enabling increased carotenogenesis. Apparently, this concept is based on the assumption that a "free carotenoid pool" controls carotenoid production by a negative feedback regulation.

An action spectrum has been elaborated by Claes[31] for the light-dependent biosynthesis of cyclic carotenes in the mutant 5/520 of *Chlorella vulgaris*. Red light near 670 nm proved to be most effective and a minor maximum was found in the blue region. Since no inhibition of the red light effect by simultaneous illumination with far-red light could be observed, the data were taken for evidence that chlorophylls are the photoreceptors in this organism.

In *Euglena,* besides the photoreceptor inside the plastids, which is most likely proto-chlorophyll(ide), an additional photoreceptor outside the plastids with high absorption in the blue region of the spectrum has been assumed.[53] Whether this is a photoprocess different from that discussed in Section III.B.1.b needs further clarification.

3. Porphyrins

Action spectra of carotenoid formation in *Mycobacterium marinum*[54] and *Myxococcus xanthus*[55] have been found to be similar; these bacteria exhibit strict photoregulation of carotenogenesis. From the shape of these spectra (Figure 6), and by comparison with absorption spectra of porphyrin-containing fractions of bacterial cell homogenates, it seems very likely that a porphyrin is the acting photoreceptor. In *Mycobacterium marinum,* mesoporphyrin or coproporphyrin, and in *Myxococcus xanthus* protoporphyrin are the favorite candidates.

Using the fungal species *Leptosphaeria michotii,* in which light increases carotenoid production, Jerebzoff-Quintin et al.[56] obtained an action spectrum with 5 peaks at 405, 480, 510, 610, and 660 nm, and also suggested a porphyrin to be the photoreceptor responsible for the reaction.

*4. UV-Light Receptor(s)**

For the fungus *Verticillium agaricinum,* Valadon and co-workers[57] reported maximum effects on the photoregulation of carotenogenesis after continuous irradiation with light of the near-UV region (peak at 370 nm), whereas blue light (380 to 525 nm, peak at 475 nm) and other spectral regions produced only very weak effects (Figure 6). Although the spectral effectiveness is not unlike that in *Mycobacterium marinum,* a yet unknown photoreceptor pigment was suggested since no porphyrins have been identified in the mycelium of this fungus. Unpublished observations by Arpin (quoted by Osman and Valadon[57]) indicating that cultures of *Pyronema omphaloides* are strongly colored by carotenoids only in near-UV light point to a similar situation. However, whether a photoreceptor different from those so far described is responsible for photoregulation in these fungi needs further investigation.

The action spectrum for *Neurospora* carotenogensis shows, besides the action in the blue, a peak in the UV-A region (and this photoresponse therefore was classified as cryptochromal), but it also shows action in the UV-B.[44] Similar characteristics were reported on carotenogenesis in *Verticillium.*[58] The UV-B peak might be due to fluorescence (as, of course, the UV-A peak) but might also be the manifestation of a different photoreceptor pigment or system. A UV-B photoreceptor was reported for some development processes (cf. Reference 59; see also Reference 60). The question of whether two different photoreceptors may be involved in carotenogenesis in *Neurospora* remains to be solved. Dual control is a widespread phenomenon in the case of blue/UV light and phytochrome.[19,59]

B. Levels of Control

Considering the biochemical background of all known photoregulations of biosynthetic processes, in particular the molecular machinery for the formation of enzymes, in principle there are four possible levels for the action of light (Figure 7):

1.   A direct transformation of a light-absorbing compound — a certain carotenoid — which then facilitates subsequent biosynthetic steps.

---

*   UV is defined as <400 nm with subsections denoted as UV-A (320 to 400 nm) UV-B (280 to 320 nm), and UV-C (<280 nm).

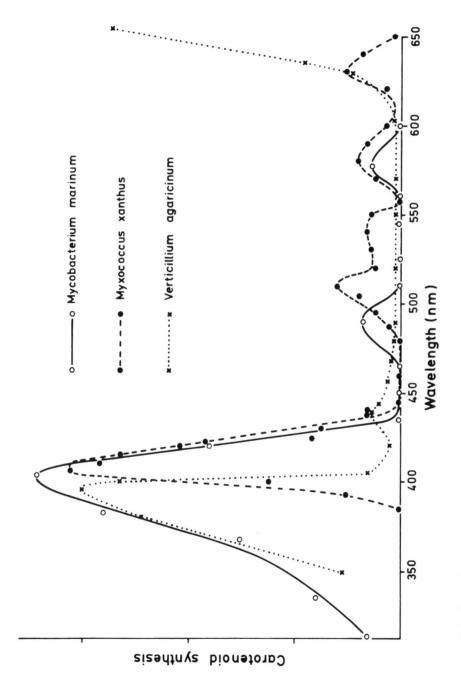

FIGURE 6. Porphyrin type action spectra of photoinduced carotenoid biosynthesis in *M. marinum* (Batra, P. P. and Rilling. H. C.[54]), *M. xanthus* (Burchard, R. P. and Hendricks, S. B.[55]) and *V. agaricinum* (Osman, M. and Valadon, L. R. G.[57]).

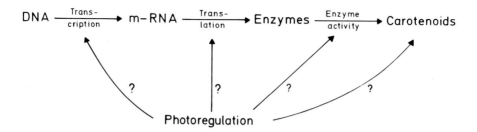

FIGURE 7.   Possible levels for the action of light in photoregulation of carotenoid biosynthesis.

2.   Photoinduced changes in the activity of carotenogenic enzymes already present.
3.   *De novo* synthesis of carotenogenic enzymes from messenger-RNA already present by regulation of translation.
4.   *De novo* synthesis of the enzymes via regulation of transcription; that is, *de novo* synthesis of the appropriate messenger-RNAs.

For each of the four possible levels a well-documented example of a photoregulation different to photoinduced carotenogenesis is known.

1. Algae
a. Biosynthetic Steps under Photocontrol

As mentioned before, conclusive results have been obtained only from mutants of the green algae *Chlorella vulgaris* and *Scenedesmus obliquus*. They clearly show that dark-grown cells of all mutant strains are unable to synthesize cyclic carotenes and xanthophylls in amounts normally present in cultures of the wild types from which the mutants had been derived. However, the presence of trace or appreciable amounts of these compounds in three of the mutants indicate that they are ''leaky'', to some extent. Neglecting this leakage, the data suggest that the photoregulated biosynthetic step may be the cyclization of carotenes. This has been put forward by Claes[30,31] for the situation in the mutant 5/520 of *Chlorella;* she assumed in addition that the cyclization is a direct photochemical reaction sensitized by chlorophylls in the light.

A more close and critical view of the data reveals that the assumption of a photoregulated cyclization reaction is premature. If in dark-grown cultures only cyclization were blocked, one would expect the accumulation of lycopene, analogous to the car R mutation of *Phycomyces blakesleeanus* (see Reference 61). However, in all three strains for which quantitative data are available, the predominant compound is ζ-carotene, whereas neurosporene and lycopene are present in minor amounts or are completely absent. To reconcile this with photoregulation of cyclization one must assume a feedback inhibition by lycopene of its own biosynthesis, or a similar mechanism. Inhibition of several steps in the biosynthetic pathway of carotenoids by feedback mechanism has already been discussed by Davies.[4]

b. Mechanisms of Photoregulation

From the data described previously, one might suggest that in the dark only enzymes for the formation of acyclic carotenoids are present (that means such carotenoids are formed independently of light), and illumination triggers the formation of cyclases and hydroxylases. However, kinetics of the changes in pigment composition during illumination in a mutant of *Scenedesmus*[35] show a rapid fall in phytoene and ζ-carotene concentrations immediately after the onset of light, that is without any lag period. Furthermore, they also show a concomitant and immediate increase of β-carotene and xanthophylls. Similar results have been obtained in studies by Senger and Strassberger,[37] who used a different mutant strain of *Scenedesmus*, as well as by Claes[30] with *Chlorella* mutants.

In summary, a common characteristic of each of the three cases is the absence of any lag phase in the photoinduced pigment transformation. Furthermore, the results obtained with *Chlorella* mutants and *Euglena* show that pigment transformation operates only during illumination. Consequently, photoinduction of *de novo* formation of the carotenogenic enzymes can very likely be ruled out. Instead, it might be considered that light acts as a "modulating" factor as in some cases of photomorphogenesis of higher plants *or*, similarly, to the photoregulation of carotenoid accumulation in higher plants. Whether such a modulation might be caused by a light-mediated change of enzyme activity as in flavin-containing enzymes as reviewed in Reference 62 is at the moment a mere speculation. In this case, a quantitative change would not be sufficient to explain the results; much more, a qualitative change has to be assumed.

In streptomycin-bleached cells of *E. gracilis* broad-band blue light (360 to 560 nm) has been reported to be most effective for stimulation of β-carotene synthesis.[26] Although in the wild type the situation seems to be more complicated, in the bleached mutant strain W$_3$BUL, that is deficient of protochlorophyllide in dark-grown cultures, a very interesting photoreaction has been detected by Schiff and co-workers. Blue light minus dark difference spectra revealed photoisomerisation of *cis-* to *trans-*ζ-carotene[63]; additional results demonstrated[64] that no other molecule but *cis-*ζ-carotene is the photoreceptor responsible for its own isomerization. It is interesting to note that also in dark-grown cultures of *Chlorella* and *Scenedesmus* mutants the acyclic carotenoids are present mainly or at least to an appreciable extent as *cis-* isomers. Therefore one might speculate that in the mutant strains mentioned, a direct *cis* to *trans* phototransformation is the rate-limiting step for subsequent biosynthesis of more unsaturated carotenes and xanthophylls. This would mean that the level of photoregulation is the compound which in turn is also the photoreceptor. However, this assumption seems to be premature since dark-grown cultures of *Scenedesmus* PG 1 and *Euglena* W$_3$BUL contain some *trans-*isomers and, moreover, in *Euglena* W$_3$BUL even cyclic compounds are present in appreciable amounts.

2. Fungi and Bacteria
a. Biosynthetic Steps under Photocontrol

In earlier investigations using *Neurospora crassa*, Zalokar[41] found that in dark-grown mycelia phytoene accumulated and colored carotenoids were synthesized only after illumination. From these and additional results,[11] it has been suggested that, in fungi, phytoene is accumulated in the dark; thus the enzyme system catalyzing the production of phytoene would have to be constitutive in dark-grown mycelia, whereas enzymes functioning in the pathway after phytoene are photoinduced.

Contrary results have been obtained from bacteria. In dark-grown cultures of *Mycobacterium sp.*, little or no phytoene is synthesized, but the phytoene content increases dramatically after illumination. Rilling and co-workers,[54] using a cell-free system derived from *Mycobacterium sp.*, were able to demonstrate the presence of geranylgeranyl-pyrophosphate synthetase (prenyl-transferase) in dark-grown cells, and a severalfold increase of the enzyme activity in illuminated cultures. Prephytoene synthetase was absent from dark-grown cells and its *de novo* synthesis was photoinduced. No significant effects of light on earlier steps in the carotenogenic pathway have been found (see Reference 11). From these results it appears fairly obvious that in bacteria (at least in *Mycobacterium sp.*) geranylgeranyl-pyrophosphate synthetase is the first enzyme of carotenoid biosynthesis that is photocontrolled. It is, however, present to some extent in dark-grown cells and hence parly constitutive; prephytoene synthetase, then, is the first enzyme the formation of which is strictly photoregulated.

In mycelia of the fungus *Fusarium aquaeductuum* the time course of incorporation of (^{14}C)-mevalonic acid into phytoene clearly indicates that there is only a very small rate of synthesis in the dark which is dramatically increased after photoinduction (Figure 8). Evi-

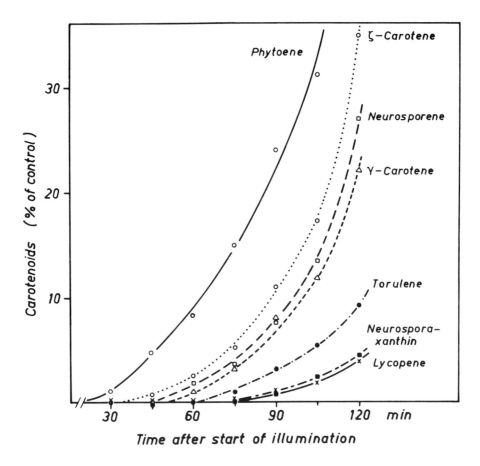

FIGURE 8.   Time course of synthesis of individual carotenoids in *Fusarium aquaeductuum* after a 10-min illumination. Synthesis of phytoene calculated from incorporation of 2-(^{14}C)-mevalonic acid (Rau, W.;[9] content of other carotenoids from Bindl, E., et al.[65])

dence that synthesis of phytoene is photoregulated has been obtained in the past years also for *Neurospora crassa*. Cell-free preparations from this fungus could be elaborated containing the phytoene-producing enzyme system thus enabling a more direct examination. An incorporation of (^{14}C)-isopentenyl-pyrophosphate into phytoene catalyzed by an enzyme preparation (100,000 × g supernatant) recovered from homogenates of mycelia was found.[66] In this system phytoene-synthesizing activity was low in preparations derived from dark-grown mycelia; but when prepared after illumination of the mycelia with blue light a ninefold increase in activity was observed. Inhibition of this photoinduced enhancement of activity with cycloheximide indicated that the enzymes responsible for the reaction may be formed *de novo*. In principle, we obtained the same results.[67]

In addition, cell-fractionation studies revealed that the phytoene-synthesizing activity was localized in a particulate cell fraction, whereas the synthesizing activity for geranylgeranyl-pyrophosphate was found to be present in the soluble fraction. Moreover, we could demonstrate a photoinduced increase in incorporating activity into more unsaturated carotenoids. Also, in our experiments, application of cycloheximide to the mycelia prior to illumination inhibited photoinduced enhancement of phytoene-synthesizing activity.

In conclusion, the data show that in fungi and nonphotosynthetic bacteria the first step(s) under photocontrol are those leading to the formation of the first C_{40}-compound, i.e., phytoene. However, it should be emphasized that photoregulation is leaky, that is, some

REFERENCES

1. **Whittingham, C. P.,** Function in photosynthesis, in *Chemistry and Biochemistry of Plant Pigments,* Goodwin, T. W., Ed., Academic Press, London, 1976, 624.
2. **Burnett, J. H.,** Functions of carotenoids other than in photosynthesis, in *Chemistry and Biochemistry of Plant Pigments,* Goodwin, T. W., Ed., Academic Press, London, 1976, 655.
3. **Krinsky, N. I.,** Carotenoid protection against oxidation, *Pure Appl. Chem.,* 51, 649, 1979.
4. **Davies, B. H.,** Carotenoid biosynthesis, in *Pigments in Plants,* 2nd ed., Czygan, F. C., Ed., Fischer, Stuttgart, 1980, 31.
5. **Goodwin, T. W.,** *The Biochemistry of the Carotenoids,* Vol. I, Chapman and Hall, London, 1980.
6. **Spurgeon, S. L. and Porter, J. W.,** Biosynthesis of carotenoids, in *Biosynthesis of Isoprenoid Compounds,* Vol. 2, Porter, J. W. and Spurgeon, S. L., Eds., John Wiley & Sons, New York, 1983.
7. **Porter, J. W. and Lincoln, R. E.,** Lycopersicon selections containing a high content of carotenes and colorless polyenes. The mechanism of carotene biosynthesis. *Arch. Biochem. Biophys.,* 27, 390, 1950.
8. **Batra, P. P.,** Mechanism of light-induced carotenoid synthesis in nonphotosynthetic plants, in *Photophysiology,* Vol. 6, Giese, A. C., Ed., Academic Press, New York, 1971, 47.
9. **Rau, W.,** Photoregulation of carotenoid biosynthesis in plants, *Pure Appl. Chem.,* 47, 237, 1976.
10. **Rau, W.,** Blue light-induced carotenoid biosynthesis in microorganisms, in *The Blue Light Syndrome,* Senger, H., Ed., Springer Verlag, Berlin, 1980, 283.
11. **Harding, R. W. and Shropshire, W., Jr.,** Photocontrol of carotenoid biosynthesis, *Annu. Rev. Plant. Physiol.,* 31, 217, 1980.
12. **Schrott, E. L.,** Carotenogenesis, in *Blue Light Effects in Biological Systems,* Senger, H., Ed., Springer Verlag, Berlin, 1984, 366.
13. **Britton, G.,** Biosynthesis of carotenoids, in *Chemistry and Biochemistry of Plant Pigments,* Vol. 1, Goodwin, T. W., Ed., Academic Press, London, 1976, 262.
14. **Thomas, R. L. and Jen, J. J.,** Phytochrome-mediated carotenoid biosynthesis in ripening tomatoes, *Plant Physiol.,* 56, 452, 1975.
15. **Simpson, D. J. and Lee, T. H.,** Chromoplast ultrastructure of *Capsicum* carotenoid mutants. II. Effect of light and CPTA, *Z. Pflanzenphysiol.,* 83, 309, 1977.
16. **Ogawa, T., Inoue, Y., Kitajima, M., and Shibata, K.,** Action spectra for biosynthesis of chlorophylls a and b and β-carotene, *Photochem. Photobiol.,* 18, 229, 1973.
17. **Frosch, S. and Mohr, H.,** Analysis of light-controlled accumulation of carotenoids in mustard (*Sinapsis alba* L.) seedlings, *Planta,* 148, 279, 1980.
18. **Mancinelli, A. L. and Rabino, I.,** The high irradiance responses of plant photomorphogenesis, *Bot. Rev.,* 44, 129, 1978.
19. **Mohr, H.,** Interaction between blue light and phytochrome in photomorphogenesis, in *The Blue Light Syndrome,* Senger, H., Ed., Springer Verlag, Berlin, 1980, 97.
20. **Goodwin, T. W.,** Distribution of carotenoids, in *Chemistry and Biochemistry of Plant Pigments,* Vol. 1, Goodwin, T. W., Ed., Academic Press, London, 1976, 225.
21. **Liaaen-Jensen, S.,** Carotenoids — a chemosystematic approach, *Pure Appl. Chem.,* 51, 661, 1979.
22. **Schiff, J. A.,** Blue light and the photocontrol of chloroplast development in *Euglena,* in *The Blue Light Syndrome,* Senger, H., Ed., Springer Verlag, Berlin, 1980, 495.
23. **Goodwin, T. W. and Jamikorn, M.,** Studies in carotenogenesis. Some observations on carotenoid synthesis in two varieties of *Euglena gracilis, J. Protozool.,* 1, 216, 1954.
24. **Wolken, J. J., Mellon, A. D., and Greenblatt, C. L.,** Environmental factors affecting growth and chlorophyll synthesis in *Euglena.* I. Physical and chemical; II. The effectiveness of the spectrum for chlorophyll synthesis, *J. Protozool.,* 2, 89, 1955.
25. **Krinsky, N. I., Gordon, A., and Stern, A. I.,** The appearance of neoxanthin during the regreening of dark-grown *Euglena, Plant Physiol.,* 29, 441, 1964.
26. **Dolphin, W. D.,** Photoinduced carotenogenesis in chlorotic *Euglena gracilis, Plant Physiol.,* 46, 685, 1970.
27. **Vaisberg, A. J. and Schiff, J. A.,** Events surrounding the early development of *Euglena* chloroplasts. VII. Inhibition of carotenoid biosynthesis by the herbicide San 9789 (4-chloro-5-methylamino)-2-(α,α,α,-trifluoro-*m*-tolyl)-3(2*H*)pyridazinone) and its developmental consequences, *Plant Physiol.,* 57, 260, 1976.
28. **Goodwin, T. W.,** Some observations on carotenoid synthesis by the alga *Chlorella vulgaris, Experientia,* 10, 213, 1954.
29. **Dresbach, C. and Kowallik, W.,** Eine fördernde Wirkung von Blaulicht auf die Carotinoidbildung einer gelben *Chlorella*-Mutante, *Planta,* 120, 291, 1974.
30. **Claes, H.,** Biosynthese von Carotinoiden bei *Chlorella.* V. Die Trennung von Licht- und Dunkelreaktionen bei der lichtabhängigen Xantophyllsynthese von *Chlorella, Z. Naturforsch.,* 14b, 4, 1959.

31. **Claes, H.,** Maximal effectiveness of 670 mμ in the light-dependent carotenoid synthesis in *Chlorella vulgaris, Photochem. Photobiol.,* 5, 515, 1966.
32. **Powls, R. and Britton, G.,** A series of mutant strains of *Scenedesmus obliquus* with abnormal carotenoid compositions, *Arch. Microbiol.,* 113, 275, 1977.
33. **Britton, G., Powls, R., and Schulze, R. M.,** The effect of illumination on the pigment composition of the ζ-carotenic mutant, PG 1, of *Scenedesmus obliquus, Arch. Microbiol.,* 113, 281, 1977.
34. **Britton, G. and Powls, R.,** Phytoene, phytofluene and ζ-carotene isomers from a *Scenedesmus obliquus* mutant, *Phytochemistry,* 16, 1253, 1977.
35. **Powls, R. and Britton, G.,** The roles of isomers of phytoene, phytofluene and ζ-carotene in carotenoid biosynthesis by a mutant strain of *Scenedesmus obliquus, Arch. Microbiol.,* 115, 175, 1977.
36. **Strassberger, G.,** Die Entwicklung des Photosyntheseapparates der Pigmentmutante C-6D von *Scenedesmus obliquus* unter besonderer Berücksichtigung der Rolle der lichtabhängig gebildeten Carotinoide, Ph.D. thesis, University of Marburg, Marburg, West Germany, 1976.
37. **Senger, H. and Strassberger, G.,** Development of the photosystems in greening algae, in *Chloroplast Development,* Akoyunoglou, G. and Argyroudi-Akoyunoglou, J. H., Eds., Elsevier, Amsterdam, 1978, 367.
38. **Rau, W.,** Photoregulation of carotenoid biosynthesis: an example of photomorphogenesis, in *Pigments in Plants,* 2nd ed., Czygan, F. C., Ed., Fischer, Stuttgart, 1980, 80.
39. **Rau, W.,** Photoregulation of carotenoid biosynthesis, in *Biosynthesis of Isoprenoid Compounds,* Vol. 2, Porter, J. W. and Spurgeon, S. L., Eds., John Wiley & Sons, New York, 1983, 123.
40. **Schrott, E. L.,** Fluence response relationship of carotenogenesis in *Neurospora crassa, Planta,* 150, 174, 1980.
41. **Zalokar, M.,** Biosynthesis of carotenoids in *Neurospora.* Action spectrum of photoactivation, *Arch. Biochem. Biophys.,* 56, 318, 1955.
42. **Rau, W.,** Untersuchungen über die lichtabhängige Carotinoidsynthese. I. Das Wirkungsspektrum von *Fusarium aquaeductuum, Planta,* 72, 14, 1967.
43. **Howes, C. D. and Batra, P. P.,** Mechanism of photoinduced carotenoid synthesis: further studies on the action spectrum and other aspects of carotenogenesis, *Arch. Biochem. Biophys.,* 137, 175, 1970.
44. **DeFabo, E. C., Harding, R. W., and Shropshire, W., Jr.,** Action spectrum between 260 and 800 nanometers for the photoinduction of carotenoid biosynthesis in *Neurospora crassa, Plant Physiol.,* 57, 440, 1976.
45. **Jayaram, M., Presti, D., and Delbrück, M.,** Light-induced carotene synthesis in *Phycomyces, Exp. Mycol.,* 3, 42, 1979.
46. **Rau, W.,** Untersuchungen über die lichtabhängige Carotinoidsynthese. IV. Die Rolle des Sauerstoffs bei der Lichtinduktion, *Planta,* 84, 30, 1969.
47. **Theimer, R. R. and Rau, W.,** Untersuchungen über die lichtabhängige Carotinoidsynthese. V. Aufhebung der Lichtinduktion durch Reduktionsmittel und Ersatz des Lichtes durch Wasserstoffperoxid, *Planta,* 92, 129, 1970.
48. **Rau, W. and Rau-Hund, A.,** Light-dependent carotenoid synthesis. X. Lag-phase after a second illumination period in *Fusarium aquaeductuum* and *Neurospora crassa, Planta,* 136, 49, 1977.
49. **Lang-Feulner, J. and Rau, W.,** Redox dyes as artificial photoreceptors in light-dependent carotenoid synthesis, *Photochem. Photobiol.,* 21, 179, 1975.
50. **Britz, S. J., Schrott, E. L., Widell, S., and Briggs, W. R.,** Red light-induced reduction of a particle-associated b-type cytochrome from corn in the presence of methylene blue, *Photochem. Photobiol.,* 29, 259, 1979.
51. **Senger, H.,** Ed., *The Blue Light Syndrome,* Springer Verlag, Berlin, 1980.
52. **Wolken, J. J. and Mellon, A. D.,** The relationship between chlorophyll and the carotenoids in the algal flagellate *Euglena, J. Gen. Physiol.,* 39, 675, 1956.
53. **Egan, J. M., Jr., Dorsky, D., and Schiff, J. A.,** Events surrounding the early development of *Euglena* chloroplasts. VI. Action spectra for the formation of chlorophyll, lag elimination in chlorophyll synthesis, and appearance of IPN-dependent triose phosphate dehydrogenase and alkaline DNase activities, *Plant Physiol.,* 56, 318, 1975.
54. **Batra, P. P. and Rilling, H. C.,** On the mechanism of photoinduced carotenoid synthesis: aspects of the photoinductive reaction, *Arch. Biochem. Biophys.,* 107, 485, 1964.
55. **Burchard, R. P. and Hendricks, S. B.,** Action spectrum for carotenogenesis in *Myxococcus xanthus, J. Bacteriol.,* 97, 1165, 1969.
56. **Jerebzoff-Quintin, S., Jerebzoff, S., and Jacques, R.,** Caroténogenese et rythme endogène de sporulation chez le *Leptosphaeria michotii* (West) Sacc. I. Action d'éclairments monochromatiques, de la diphénylamine de l''antimycine A sur l'évolution de la caroténogenèse, *Physiol. Veg.,* 13, 55, 1975.
57. **Osman, M. and Valadon, L. R. G.,** Effects of light quality on the photoinduction of carotenoid synthesis in *Verticillium agaricinum, Microbios,* 18, 229, 1977.

58. **Hsiao, K. C. and Björn, L. O.,** Aspects of photoinduction of carotenoid biosynthesis in *Verticillium agaricinum, Physiol. Plant.,* 54, 235, 1982.

59. **Wellmann, E.,** UV radiation in photomorphogenesis, in *Encyclopedia of Plant Physiology, New Series,* Vol. 16B, Shropshire, W., Jr. and Mohr, H., Eds., Springer Verlag, Berlin, 1983, 745.

60. **Drumm-Herrel, H. and Mohr, H.,** A novel effect of UV-B in a higher plant *(Sorghum vulgare), Photochem. Photobiol.,* 33, 391, 1981.

61. **Cerdá-Olmedo, E. and Torres-Martinez, S.,** Genetics and regulation of carotene biosynthesis, *Pure Appl. Chem.,* 51, 631, 1979.

62. **Zumpft, W. R., Castillo, F., and Hartmann, K. M.,** Flavin mediated photoreduction of nitrate reductase of higher plants and microorganisms, in *The Blue Light Syndrome,* Senger, H., Ed., Springer Verlag, Berlin, 1980, 422.

63. **Fong, F. and Schiff, J. A.,** Blue-light-induced absorbance changes associated with carotenoids in *Euglena, Planta,* 146, 119, 1979.

64. **Steinitz, Y. L., Schiff, J. A., Osafune, T., and Green, M. S.,** Cis to trans photoisomerisation of ζ-carotene in *Euglena gracilis* var. *bacillaris* W₃BUL: further purification and characterization of the photoactivity, in *The Blue Light Syndrome,* Senger, H., Ed., Springer Verlag, Berlin, 1980, 269.

65. **Bindl, E., Lang, W., and Rau, W.,** Untersuchungen über die lichtabhängige Carotinoidsynthese. VI. Zeitlicher Verlauf der Synthese der einzelnen Carotinoide bei *Fusarium aquaeductuum* unter verschiedenen Induktionsbedingungen, *Planta,* 94, 156, 1970.

66. **Spurgeon, S. L., Turner, R. V., and Harding, R. W.,** Biosynthesis of phytoene from isopentenyl pyrophosphate by a *Neurospora crassa* enzyme system, *Arch. Biochem. Biophys.,* 195, 23, 1979.

67. **Mitzka-Schnabel, U. and Rau, W.,** Subcellular site of carotenoid biosynthesis in *Neurospora crassa, Phytochemistry,* 20, 63, 1980.

68. **Murillo, F. J. and Cerdá-Olmedo, E.,** Regulation of carotene synthesis in *Phycomyces, Mol. Gen. Genet.,* 148, 19, 1976.

69. **Theimer, R. R. and Rau, W.,** Mutants of *Fusarium aquaeductuum* lacking photoregulation of carotenoid synthesis, *Biochim. Biophys. Acta,* 177, 180, 1969.

69a. **Rau, W.,** unpublished results, 1984.

70. **Rau, W., Feuser, B., and Rau-Hund, A.,** Substitution of p-chloro-hydroxymercuribenzoate for light in carotenoid synthesis by *Fusarium aquaeductuum, Biochim. Biophys. Acta,* 136, 589, 1976.

71. **Rau, W.,** Untersuchungen über die lichtabhängige Carotinoid-synthese. II. Ersatz der Lichtinduktion durch Mercuribenzoat, *Planta,* 74, 263, 1976.

72. **Eslava, A. P., Alvarez, M. I., and Cerdá-Olmedo, E.,** Regulation of carotene biosynthesis in *Phycomyces* by vitamin A and β-ionone, *Eur. J. Biochem.,* 48, 617, 1974.

73. **Theimer, R. R. and Rau, W.,** Untersuchungen über die lichtabhängige Carotinoidsynthese. VIII. Die unterschiedlichen Wirkungsmechanismen von Licht und Mercuribenzoat, *Planta,* 106, 331, 1972.

74. **Rau, W.,** Untersuchungen über die lichtabhängige Carotinoid-synthese. VII. Reversible Unterbrechung der Reaktionskette durch Cycloheximid und anaerobe Bedingungen, *Planta,* 101, 251, 1971.

75. **Schrott, E. L. and Rau, W.,** Evidence for a photoinduced synthesis of poly(A) containing mRNA in *Fusarium aquaeductuum, Planta,* 136, 45, 1977.

76. **Mitzka-Schnabel, U., Warm, E., and Rau, W.,** Light-induced changes in the protein pattern translated in vivo and in vitro accompanying carotenogenesis in *Neurospora crassa* and *Fusarium aquaeductuum,* in *Blue Light Effects in Biological Systems,* Senger, H., Ed., Springer Verlag, Berlin, 1984, 264.

77. **Krinsky, N. I.,** Photosensitization and singlet oxygen damage, in *Topics in Photobiology,* Kim, H.-O. and Song, P.-S., Eds., Jeju National University, Jeju, 1983, 67.

78. **Schrott, E. L.,** Carotenoids in plant photoprotection, *Pure Appl. Chem.,* 57, 729, 1985.

79. **Ridley, S. M.,** Carotenoids and herbicide ation, in *Carotenoid Chemistry and Biochemistry,* Britton, G. and Goodwin, T. W., Eds., Pergamon Press, New York, 1982, 353.

80. **Mathis, P. and Schenck, C. C.,** The functions of carotenoids in photosynthesis, in *Carotenoid Chemistry and Biochemistry,* Britton, G. and Goodwin, T. W., Eds., Pergamon Press, New York, 1982, 339.

81. **Webb, R. B.,** Lethal mutagenic effects of near-ultraviolet radiation, in *Photochemical and Photobiological Reviews,* Vol. 2, Smith, K. C., Ed., Plenum Press, New York, 1977, 169.

82. **Generoso, W. M., Shelby, M. D., and de Serres, F. J.,** *DNA Repair and Mutagenesis,* Plenum Press, New York, 1980.

83. **Schrott, E. L.,** The biphasic fluence response of carotenogenesis in *Neurospora crassa:* temporary insensitivity of the photoreceptor system, *Planta,* 151, 371, 1981.

Chapter 6

BLUE LIGHT CONTROL OF PIGMENT BIOSYNTHESIS — ANTHOCYANIN BIOSYNTHESIS

Helga Drumm-Herrel

TABLE OF CONTENTS

I. INTRODUCTION

Anthocyanins (glycosylated anthocyanidins) belong to a class of flavonoids which is almost ubiquitous in plants except in algae, fungi, lichens, and bryophytes. Anthocyanins are water-soluble, vacuolar pigments responsible for the conspicuous violet/blue and red/purple coloration of flowers, fruits, vegetative parts (including fall coloration), and young seedlings of higher plants.[1] The only exceptions known are the Centrospermae and Cactaceae where the violet/red colors are due to a different group of water-soluble, vacuolar pigments, the betacyanins.[2] Higher plants differ, of course, genetically with regard to their potential to synthesize anthocyanins.[5] Moreover, the extent of pigment synthesis and accumulation is influenced by many environmental variables, such as nutrients, temperature, availability of water, infections, and — in particular — light.[3] The chemistry, biochemistry, natural occurrence, inheritance, taxonomy, and function of anthocyanins have been studied in great detail. Several excellent and comprehensive reviews of these aspects are available.[4-11]

This chapter deals specifically with the action of light — in particular, blue light –– on anthocyanin biosynthesis.

II. BIOSYNTHESIS OF ANTHOCYANINS

The elucidation of the flavonoid pathway is based on four general approaches: the use of isotopically labeled precursors,[12,13] the investigation of enzyme systems in cell-free extracts,[14] the use of organisms with genetic blocks,[15-17] and the application of metabolic inhibitors such as aminooxyacetic acid, a potent inhibitor of anthocyanin synthesis.[18,19]

The enzymatic steps involved in the biosynthesis of flavonoids have recently been summarized in several reviews.[5,20-22] The scheme on biosynthesis of anthocyanins in Figure 1 is based on the most recent reports in this field.[5,23,16]

The first steps in flavonoid (including anthocyanin) biosynthesis are firmly established. All flavonoids derive their carbon skeleton from compounds of intermediary cell metabolism through the action of two consecutive pathways: the general phenylpropanoid and the flavonoid pathway. The enzymes catalyzing the sequence of reactions converting phenylalanine into the CoA esters are all known. They are listed in Table 1 (general phenylpropanoid metabolism[24]). The condensation of the acyl residues from one molecule of 4-coumaroyl-CoA leading to a chalcone is the central reaction of flavonoid biosynthesis. The enzyme catalyzing this step, chalcone synthase, can be regarded as a key enzyme of flavonoid biosynthesis.[5,6] Strong support for the role of dihydroflavonols as intermediates in anthocyanidin biosynthesis came from tracer studies and supplementation experiments performed with flowers of genetically defined acyanic lines.[15,25] Only very recently sound experimental evidence for the role of flavon-3,4-diols (leucoanthocyanidins) as intermediates in anthocyanin biosynthesis was obtained (see References 23 and 16). Enzyme(s) which catalyze the conversion of leucoanthocyanidins to anthocyanidins are not yet known. Nearly 40 different types of anthocyanin glycosides are known,[7,11] with cyanidin being the most abundant aglycone in vegetative organs. The aglycones occur in vivo as 3-glycosides, 3,5-bisglycosides, or 3,7-bisglycosides. The glycosides are often further substituted by acyl residues. The enzymatic reactions known so far are discussed by Ebel and Hahlbrock.[5]

III. BLUE-LIGHT EFFECTS ON ANTHOCYANIN BIOSYNTHESIS

Reports on photocontrol by blue light of anthocyanin production are scarce in the literature. Those investigators who have mainly elucidated the biosynthetic pathway of flavonoids used in general white fluorescent light to induce the enzymes and end products of this pathway. Researchers more interested in the transduction of the light signal to the final response

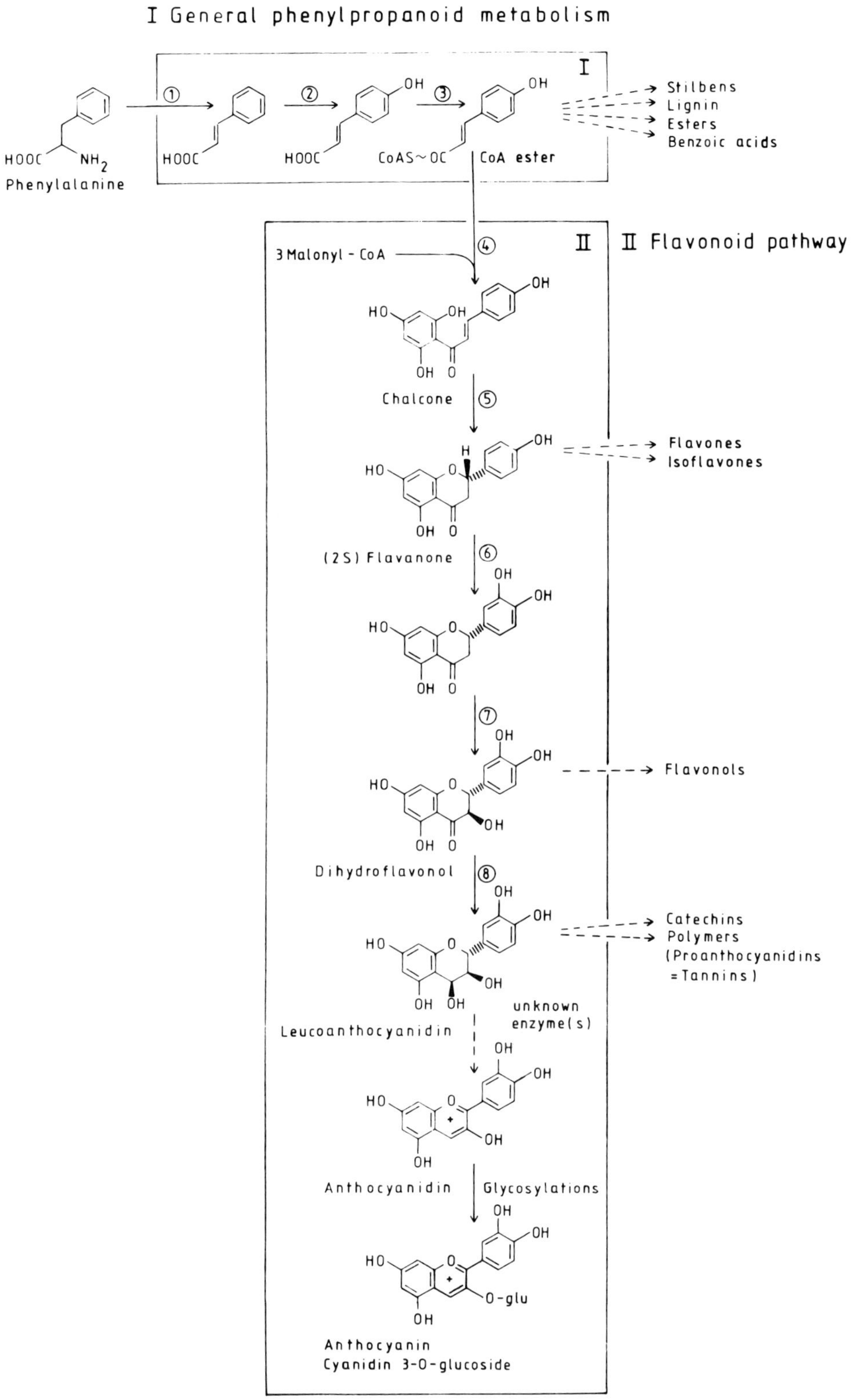

I General phenylpropanoid metabolism

FIGURE 1.   Scheme illustrating the sequence of reactions leading finally to anthocyanins. The enzymes marked by numbers are listed in Table 1.

Table 1
ENZYMES MENTIONED IN FIGURE 1

Enzyme	EC number	Key to figure (no.)
General phenylpropanoid metabolism		
Phenylalanine ammonia lyase	4.3.1.5	1
Cinnamate 4-hydroxylase	1.14.13.11	2
4-Coumarate: CoA ligase	6.2.1.12	3
Flavonoid pathway		
Chalcone synthase	—	4
Chalcone isomerase	5.5.1.6	5
"Flavanone 3'-hydroxylase"	—	6
"Flavanone 3-hydroxylase"	—	7
"Dihydroflavonol 4-reductase"	—	8

Table 2
INDUCTION (OR LACK OF INDUCTION) OF ANTHOCYANIN SYNTHESIS IN THE MESOCOTYL OF MILO SEEDLINGS *(SORGHUM VULGARE)* BY LIGHT OF DIFFERENT QUALITIES

Treatment (onset 60 hr after sowing)	Amt of anthocyanin (measurement: 87 hr after sowing; A at 510 nm)
27 hr dark	0
27 hr WL[a]	1.85
27 hr RL	0
27 hr FR	0
3 hr WL	0.19
3 hr BL/UV	0.19
3 hr WL + 5 min RL[b]	0.19
3 hr WL + 5 min 756 nm light[b]	0.06
3 hr WL + 5 min 756 nm light + 5 min RL	0.20
3 hr BL/UV + 5 min RL	0.19
3 hr BL/UV + 5 min 756 nm light	0.05
3 hr BL/UV + 5 min 756 nm light + 5 min RL	0.19

Note: In the case of a 3-hr light treatment, seedlings were kept in the dark for 24 hr before extraction of anthocyanin.

[a]   WL: xenon arc light, similar to sunlight, 250 W · m^{-2}.

[b]   Photoequilibria of the phytochrome system are of the order of $\phi_{RL} = 0.8$ and $\phi_{756} < 0.01$.[28] A 5-min light pulse suffices under the present circumstances to virtually establish the photoequilibrium $\quad \varphi_\lambda = \dfrac{[P_{fr}]_\lambda^b}{[P_{tot}]}$

worked with better defined light qualities but have not yet reached the molecular level. Therefore, only some thoroughly investigated case studies will be presented, including reports on photoregulation of flavonoids other than anthocyanins because of the close biosynthetic relations between the different subgroups of flavonoids (see Figure 1).

A. Blue Light Effects on Anthocyanin Biosynthesis in Whole Plants
1. The Milo Seedling (Sorghum vulgare *Pers. cv. Weider [hybrid]*)
The mesocotyl of the milo seedling does not produce anthocyanin in complete darkness, while white light causes a rapid pigmentation.[26] The action of white light can be replaced by irradiation with blue/ultraviolet light (BL/UV).[27] The expression of the BL/UV effect is controlled by phytochrome* (Table 2). Dichromatic irradiation with two kinds of light to

---

* P_{fr} is the far-red absorbing, physiologically active form of the phytochrome system. Some basic properties of phytochrome are described elsewhere in this volume (see Chapter 10).

Table 3
THE EFFECT OF SIMULTANEOUS IRRADIATION WITH TWO DIFFERENT LIGHT QUALITIES (BL/UV AND RG 9-LIGHT), ON ANTHOCYANIN FORMATION IN THE MESOCOTYL OF MILO SEEDLINGS

Treatment after sowing	Amt of anthocyanin (A at 510 nm)
60 hr D^a + 3 hr BL/UV (1.5 Wm^{-2}) + 5 min RL + 24 hr D	0.189
60 hr D + 3 hr [BL/UV + RG9 − light (200 Wm^{-2})] + 24 hr D	0.055
60 hr D + 3 hr [BL/UV + RG9 − light (200 Wm^{-2})] + 5 min RL + 24 hr D	0.183

Note: The intense RG9-light was applied to decrease the level of P_{fr} as far as possible. Five minutes of RL suffice to establish the maximum P_{fr}/P_{tot} ratio. D = dark.

strongly reduce the level of P_{fr} on a constant background of BL/UV shows that the BL/UV photoreaction as such is not affected by the presence or virtual absence of P_{fr} (Table 3). With and without P_{fr} during the 3-hr irradiation period, the same amount of anthocyanin is synthesized as long as a red light (RL) pulse saturating the phytochrome photoequilibrium is given after the irradiation. The present interpretation is that phytochrome (P_{fr}) is the effector molecule which causes anthocyanin synthesis through activation of competent genes (for a full argument see Mohr and Schäfer[29] and Oelmüller and Mohr[30]). With respect to anthocyanin synthesis in the milo mesocotyl, there is no detectable responsivity toward P_{fr} without the operation of a BL/UV photoreceptor, which is considered to *establish* responsivity toward phytochrome. Recently this was unambiguously demonstrated by Oelmüller and Mohr.[44]

Besides the described coaction of a BL/UV photoreceptor and phytochrome, a strong positive synergistic effect of UV-B light (280 to 320 nm) was found, if it is given simultaneously with or after blue or UV-A light.[31] This UV-B effect, probably due to a separate UV-B photoreceptor, can only express itself in the presence of phytochrome.[27]

In the same system, regulation by light of chalcone synthase, a key enzyme of flavonoid biosynthesis, is presently under study by Oelmüller (Figure 2). RL has only a very weak effect compared to UV. Using cut-off filters it can be shown that, beside UV-A (320 to 390 nm), UV-B (280 to 320 nm) is involved in this light induction. The lag phase of anthocyanin production is somewhat longer than 3 hr (Figure 2), whereas chalcone synthase activity already shows a significant increase after 3 hr of UV and RL. These data make it very likely that this enzyme indeed plays the role of a key enzyme in flavonoid and, in this particular case, in anthocyanin synthesis, as stated by Hahlbrock.[6] However, in contrast to the end product, anthocyanin there is a weak effect of RL acting through phytochrome which cannot be neglected and requires further investigation.

2. The Maize Seedling (Zea mays cv. Pioneer 3369 A)

Duke and Naylor[32] have studied the photoinduction of anthocyanin synthesis in root and mesocotyl of maize. They found that continuous white or blue light induced large amounts of anthocyanin in both organs, whereas continuous far-red (FR) operating through phytochrome induced little pigmentation. However, after a white light (WL) pretreatment, the response is under phytochrome control. If the WL irradiation is terminated by a 5-min FR pulse to return P_{fr} to P_r, the final amount of anthocyanin is greatly decreased, and this effect can be reversed by giving a RL pulse after the FR pulse. With respect to anthocyanin synthesis, it seems that the maize system acts very similar to the milo system, except that it responds weakly to P_{fr} without a BL or WL pretreatment. In this case a WL or BL pretreatment causes a strong "responsivity amplification" for P_{fr}[30] rather than establishing responsivity for P_{fr} as in the milo seedling.

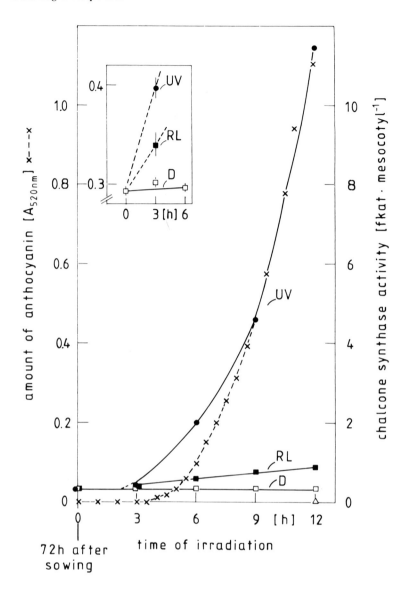

FIGURE 2.   Time courses of anthocyanin accumulation and chalcone synthase activity in the mesocotyl of 72-hr-old dark-grown milo seedlings irradiated with RL (6.7 Wm^{-2}) or UV (9.6 Wm^{-2}). Anthocyanin is completely absent in dark (D) or RL ($\triangle$). The inset shows that after 3 hr of dark, irradiation with RL or UV chalcone synthase activity is already significantly higher than the D level.[45]

3. The Tomato Seedling (Lycopersicon esculentum *Mill., cv. Hilds Matina*)

Single light pulses of different light quality have no detectable effect on the production of anthocyanin in the hypocotyl of tomato seedlings even though long-term exposure to WL, BL, or UV leads to strong anthocyanin synthesis.[33] The effect of high fluence rate RL or FR is only small compared to WL. It was previously shown by Mancinelli and Rabino[34] that the long-term RL and FR treatment was exclusively mediated through phytochrome.

As in the maize seedling, anthocyanin synthesis after BL or UV exposure continues in darkness at a considerable rate only if P_{fr} is available. When most of the P_{fr} is returned to P_r by a saturating RG9-light pulse, further anthocyanin synthesis is almost negligible (Figure 3). Even though phytochrome is involved in the response, a high effectiveness of phyto-

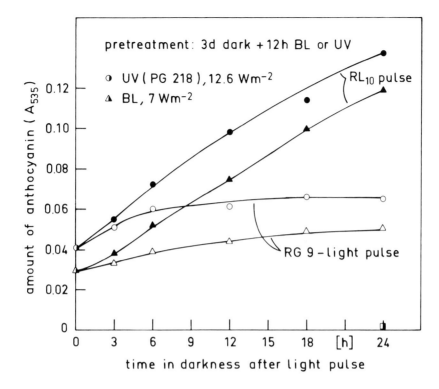

FIGURE 3. Anthocyanin accumulation in the tomato hypocotyl in darkness following a light treatment. The seedlings were pretreated either with UV or with BL. The pretreatment was terminated either with a saturating RL pulse (RL_{10}, $\phi_{RL} = 0.8$, see Table 2) or with a saturating RG 9-light pulse ($\phi_{RG\ 9} = \phi_{756} < 0.01$, see Table 2).[33]

chrome depends on a preceding light absorption by a BL/UV photoreceptor. Our present interpretation is that, as in the maize seedling, a short wavelength pretreatment leads to "responsivity amplification" for P_{fr}.

B. Blue-Light Effects on Anthocyanin Biosynthesis in Cell Cultures

1. Cell Cultures of Haplopappus gracilis Cass.

Lackmann[35] demonstrated that in callus cultures as well as in intact seedlings of *Haplopappus gracilis* anthocyanin synthesis is induced by BL and UV-A. Action spectra show only two peaks at 438 and 372 nm and no peak in the red region of the spectrum. On the other hand, Wellmann et al.,[36] working with the same system, demonstrated that only long-term irradiations of UV-A (320 to 390 nm) *plus* UV-B (280 to 320 nm) at a relatively high fluence rate were effective in inducing anthocyanin and enzymes related to its biosynthesis. The activities of phenylalanine ammonia lyase, chalcone synthase, and chalcone isomerase increase simultaneously with anthocyanin accumulation. From the data available at present, an involvement of phytochrome cannot be completely excluded.

2. Cell Cultures of Parsley (Petroselinum hortense Hoffm.)

With the aid of cell cultures of parsley the details of the biosynthesis of flavonoids have been elucidated to a large extent.[6] Irradiation of dark-grown cell suspension cultures causes a coordinated and selective induction of about 16 enzymes, all of which are involved in the formation of several flavonoid glycosides which accumulate in irradiated cells.[21,37]

Wellmann[38,39] and Wellmann and Schopfer[40] have shown that in this system the induction has an absolute requirement for UV-B light, and that the expression is under phytochrome control. On the other hand Duell-Pfaff and Wellmann[41] have found that irradiations with red, far-red, and blue light *before* and *after* the UV stimulus increase the amount of flavonoids, with blue light being the most effective spectral range. The authors imply that in parsley cell cultures, beside the action of UV-B and phytochrome, a separate blue light photoreceptor is involved. This statement needs further corroberation.

In the same system Kreuzaler et al.[42] have shown that UV-induced changes in the amount of chalcone synthase mRNA coincide with the UV-induced changes of chalcone synthase synthesis in vivo and in vitro.

Recently, Chappell and Hahlbrock[43] found transiently increased transcription rates of the often claimed hypothesis that UV induction of these enzymes at important positions of the flavonoid pathway acts at the level of gene expression.
phenylalanine ammonia lyase and chalcone synthase genes in nuclei isolated from parsley cell cultures after a 2.5-hr irradiation with UV. Taking these data together, they verify the

IV. CONCLUSIONS

The case studies presented lend support to a unifying concept. At least in the case of whole plants a *coaction* of BL/UV and light absorbed by phytochrome is observed: phytochrome (P_{fr}) is hardly or not at all capable of operating without a preceding treatment with BL/UV, while the action of BL/UV requires P_{fr} for expression. However, the different species differ greatly with regard to the RL-, BL-, and UV-dependent processes they must perform in order to establish responsivity towards phytochrome (P_{fr}) or to amplify responsivity toward P_{fr}.

In the case of cell cultures, the situation is not clear as far as the photoreceptor is concerned. But at least in *Haplopappus* cultures, a BL/UV photoreceptor in addition to a UV-B photoreceptor seems to be involved. It might turn out after further investigation that the above explanation of phenomena (sequential action of BL/UV light and light absorbed by phytochrome) is applicable to cell cultures as well. In any case, the data from cell cultures also suggest that a P_{fr}-mediated modulation of activity of competent genes leading finally to flavonoid or specifically anthocyanin synthesis is the core of the light response.

Coaction of different photoreceptors is a more general phenomenon in plant photomorphogenesis. The *mode* of coaction will be described by H. Mohr in Chapter 4.

ACKNOWLEDGMENTS

The research was supported by the Deutsche Forschungsgemeinschaft (SFB 206). I am greatly indebted to R. Oelmüller for fruitful discussions.

REFERENCES

1. **Swain, T.,** Nature and properties of flavonoids, in *Chemistry and Biochemistry of Plant Pigments,* Goodwin, T. W., Ed., Academic Press, New York, 1976, 425.
2. **Piattelli, M.,** The betalains: structure, biosynthesis and chemical taxonomy, in *The Biochemistry of Plants, a Comprehensive Treatease,* Vol. 7, Stumpf, P. K. and Conn, E. E., Eds., Academic Press, New York, 1981, 557.
3. **Mancinelli, A. L.,** The photoregulation of anthocyanin synthesis, in *Encyclopedia of Plant Physiology,* New Series, Vol. 16B, Shropshire, W., Jr. and Mohr, H., Eds., Springer-Verlag, Berlin, 1983.
4. **Bell, E. A.,** The physiological role(s) of secondary (natural) products, in *The Biochemistry of Plants, a Comprehensive Treatease,* Vol. 7, Stumpf, P. K. and Conn, E. E., Eds., Academic Press, New York, 1981, 1.

5. **Ebel, J. and Hahlbrock, K.,** Biosynthesis, in *The Flavonoids, Advances in Research,* Harborne, J. B. and Mabry, T. J., Eds., Chapman and Hall, London, 1982, 641.

6. **Hahlbrock, K.,** Flavonoids, in *The Biochemistry of Plants, a Comprehensive Treatease,* Vol. 7, Stumpf, P. K. and Conn, E. E., Eds., Academic Press, New York, 1981, 425.

7. **Harborne, J. B.,** *Comparative Biochemistry of the Flavonoids,* Academic Press, New York, 1967.

8. **Harborne, J. B.,** Functions of flavonoids in plants, in *Chemistry and Biochemistry of Plant Pigments,* Goodwin, T. W., Ed., Academic Press, New York, 1976, 736.

9. **Hess, D.,** *Biochemische Genetik,* Springer-Verlag, Berlin, 1968.

10. **Jurd, L.,** Recent progress in the chemistry of flavylium salts, in *Recent Advances in Phytochemistry,* Vol. 5, Runeckles, V. C. and Tso, T. C., Eds., Academic Press, New York, 1972, 135.

11. **Timberlake, C. F. and Bridle, P.,** The anthocyanins, in *The Flavonoids,* Harborne, J. B., Mabry, T. J., and Mabry, H., Eds., Chapman and Hall, London, 1975, 214.

12. **Grisebach, H. and Baaz, W.,** Biochemie der Flavonoide, *Naturwissenschaften,* 56, 538, 1969.

13. **Swain, T.,** Methods used in the study of biosynthesis, in *Biosynthetic Pathways in Higher Plants,* Pridham, J. B. and Swain, T., Eds., Academic Press, New York, 1965, 9.

14. **Hahlbrock, K. and Griesebach, H.,** Origin of flavonoids, in *The Flavonoids,* Harborne, J. B., Mabry, T. J., and Mabry, H. Eds., Chapman and Hall, London, 1975, 866.

15. **Forkmann, G.,** Precursors and genetic control of anthocyanin synthesis, *Planta,* 137, 159, 1977.

16. **Heller, W., Britsch, L., Forkmann, G., and Grisebach, H.,** Leucoanthocyanidins as intermediates in anthocyanidin biosynthesis in flowers of *Matthiola incana* R. Br., *Planta,* 163, 191, 1985.

17. **Kho, K. F. F.,** Conversion of hydroxylated and methylated dihydroflavonols into anthocyanins in a white flowering mutant of *Petunia hybrida, Phytochemistry,* 17, 245, 1978.

18. **Amrhein, N.,** Biosynthesis of cyanidin in buckwheat hypocotyls, *Phytochemistry,* 18, 585, 1979.

19. **Scherf, H. and Zenk, M. H.,** Der Einfluss des Lichtes auf die Flavonoidsynthese und die Enzyminduktion bei *Fagopyrum esculentum* Moench, *Z. Pflanzenphysiol.,* 57, 401, 1967.

20. **Grisebach, H.,** Selected topics in flavonoid biosynthesis, in *Recent Advances in Phytochemistry,* Vol. 12, Swain, T., Harborne, J. B., and Van Sumere, C. F., Eds., Plenum Press, New York, 1979, 221.

21. **Hahlbrock, K. and Grisebach, H.,** Enzymatic control in the biosynthesis of lignin and flavonoids, *Annu. Rev. Plant Physiol.,* 30, 105, 1979.

22. **Wong, E.,** Biosynthesis of flavonoids, in *Chemistry and Biochemistry of Plant Pigments,* Vol. 1, Goodwin, T. W., Ed., Academic Press, New York, 1976, 464.

23. **Grisebach, H.,** Selected topics in flavonoid biosynthesis, in *Annu. Proc. Phytochemical Soc. of Europe,* Vol. 25, Van Sumere, C. F. and Lea, P., Eds., Clarendon Press, Oxford, 1985, 183.

24. **Ebel, J., Schaller-Hekeler, B., Knobloch, K.-H., Wellmann, E., Grisebach, H., and Hahlbrock, H.,** Coordinated changes in enzyme activities of phenylpropanoid metabolism during the growth of soybean cell suspension cultures, *Biochem. Biophys. Acta,* 362, 417, 1974.

25. **Grisebach, H.,** Biosynthesis of anthocyanins in *Anthocyanins as Food Colors,* Markakis, P., Ed., Academic Press, New York, 1982, 69.

26. **Downs, R. J. and Siegelman, H. W.,** Photocontrol of anthocyanin synthesis in milo seedlings, *Plant Physiol.,* 38, 25, 1963.

27. **Drumm, H. and Mohr, H.,** The mode of interaction between blue (UV) light photoreceptors and phytochrome in anthocyanin formation of the *Sorghum* seedling, *Photochem. Photobiol.,* 27, 241, 1978.

28. **Schäfer, E., Lassig, T. U., and Schopfer, P.,** Photocontrol of phytochrome destruction in grass seedlings. The influence of wavelength and irradiance, *Photochem. Photobiol.,* 22, 193, 1975.

29. **Mohr, H. and Schäfer, E.,** Photoperception and deetiolation, *Phil. Trans. R. Soc. London Ser. B,* 303, 489, 1983.

30. **Oelmüller, R. and Mohr, H.,** Induction versus modulation in phytochrome-regulated biochemical processes, *Planta,* 161, 165, 1984.

31. **Drumm, H. and Mohr, H.,** A novel effect of UV-B in a higher plant *(Sorghum vulgare), Photochem. Photobiol.,* 33, 391, 1981.

32. **Duke, S. O. and Naylor, A. W.,** Light control of anthocyanin biosynthesis in *Zea* seedlings, *Physiol. Plant.,* 37, 62, 1976.

33. **Drumm-Herrel, H. and Mohr, H.,** The effect of prolonged light exposure on the effectiveness of phytochrome in anthocyanin synthesis in tomato seedlings, *Photochem. Photobiol.,* 35, 233, 1982.

34. **Mancinelli, A. L. and Rabino, J.,** The high irradiance responses of plant photomorphogenesis, *Bot. Rev.,* 44, 129, 1978.

35. **Lackmann, I.,** Wirkungsspektren der Anthocyansynthese in Gewebekulturen und Keimlingen von *Haplopappus gracilis, Planta,* 98, 258, 1971.

36. **Wellmann, E., Hrazdina, G., and Grisebach, H.,** Induction of anthocyanin formation and of enzymes related to its biosynthesis by UV light in cell cultures of *Haplopappus gracilis, Phytochemistry,* 15, 913, 1976.

37. **Hahlbrock, K., Knobloch, K. H., Kreuzaler, F., Potts, J. R. M., and Wellmann, E.,** Coordinated induction and subsequent activity changes of two groups of metabolically interrelated enzymes, *Eur. J. Biochem.,* 61, 199, 1976.

38. **Wellmann, E.,** Phytochrome-mediated flavone synthesis in cell suspension cultures of *Petroselinum hortense* after preirradiation with ultraviolet light, *Planta,* 101, 283, 1971.

39. **Wellmann, E.,** Specific ultraviolet effects in plant morphogenesis. Yearly review, *Photochem. Photobiol.,* 24, 659, 1976.

40. **Wellmann, E. and Schopfer, P.,** Phytochrome-mediated *de novo* synthesis of phenylalanine ammonialyase in cell suspension cultures of parsley, *Plant Physiol.,* 55, 822, 1975.

41. **Duell-Pfaff, N. and Wellmann, E.,** UV-B induced flavonoid synthesis in cell suspension cultures of parsley *(Petroselinum hortense* Hoffm.). The role of phytochrome and a blue light photoreceptor, *Planta,* 156, 213, 1982.

42. **Kreuzaler, F., Ragg, H., Fautz, E., Kuhn, D. N., and Hahlbrock, K.,** UV-induction of chalconesynthase mRNA in cell suspension cultures of *Petroselinum hortense, Proc. Natl. Acad. Sci. U.S.A.,* 80, 2591, 1983.

43. **Chappell, J. and Hahlbrock, K.,** Transcription of plant defense genes in response to UV light or fungal elicitor, *Nature (London),* 311, 76, 1984.

44. **Oelmüller, R. and Mohr, H.,** Mode of coaction between blue/UV light and light absorbed by phytochrome in light-mediated anthocyanin formation in the milo *(Sorghum vulgare* Pers.) seedling, *Proc. Natl. Acad. Sci. U.S.A.,* 82, 6124, 1985.

45. **Oelmüller, R.,** Die Abhängigkeit der Phytochromwirkung vom Lichtfaktor und von intrazellulären Signalen, Ph.D. thesis, University of Freiburg, Freiburg, West Germany, 1985.

Chapter 7

BLUE LIGHT CONTROL OF PIGMENT BIOSYNTHESIS — CHLOROPHYLL BIOSYNTHESIS

Horst Senger

TABLE OF CONTENTS

I. INTRODUCTION

The biosynthesis of chlorophyll can either be completely independent from light (green algae), light dependent (angiosperms), or light independent to a certain extent (gymnosperms). In all cases it is indispensable that precursors for the biosynthesis be provided either by photosynthesis or heterotrophic nutrition. For example, green algae form chlorophyll in the dark only if heterotrophic nutrition is provided. Photosynthesis, the classical light process, provides precursors and energy. But this is only an indirect light effect, and photosynthetic influence on chlorophyll biosynthesis is not considered here.

Various modes of light action upon chlorophyll biosynthesis are known: the direct control of the biosynthetic pathways of porphyrin and of phytol, the influence on the chlorophyll a/b ratio, and the influence on the protein formation of the pigment protein complexes. Among all these possibilities the regulatory effect of light on porphyrin biosynthesis has been studied most extensively, and the present chapter will focus on this topic.

Studies on the light induction of phytol biosynthesis revealed that it takes place in higher plants only in light, demonstrating an action spectrum similar to that of total chlorophyll formation.[1] Esterification of chlorophyllide with phytol has also been demonstrated to be light dependent.[2] On the other hand, it has been established that phytochrome, the common light receptor for photomorphogenetic effects in higher plants, is not involved in the esterification of chlorophyllide.[3] Thus the final proof that phytol formation and/or esterification of chlorophyllide are directly light dependent has still to be furnished. So far it can not be excluded that the light-dependent protochlorophyllide reduction controls either phytol formation and/or esterification of chlorophyllide via a feedback mechanism.

The influence of blue light on the chlorophyll a/b ratio, and the biosynthesis of the chlorophyll proteins will be discussed in Section III.

II. CHLOROPHYLL BIOSYNTHESIS IN ALGAE

A. Conditions of Greening

As mentioned before, algae generally have the ability to perform chlorophyll biosynthesis entirely in the dark if appropriate substrates for heterotrophic growth are provided. Nevertheless, there are two possibilities of making chlorophyll biosynthesis in algae light dependent: nonpermissive nutrient or growth conditions and induction of pigment mutations.

In several green algae, nitrogen deficiency or suboptimal temperatures cause a depression of chlorophyll biosynthesis in the dark (Table 1). Regreening of nitrogen-starved cells is completely light dependent, whereas the cells grown under nonpermissive temperatures can form a certain amount of chlorophyll in the dark when transferred to optimal growth temperatures. Complete greening of the temperature mutants is only accomplished in light. The same light dependence of greening becomes visible in mutants, which either occur spontaneously or are induced by UV irradiation. Regarding the regulation of pigment biosynthesis, the genetic material appears to be specifically sensitive to this type of mutation, since it occurs quite frequently in various organisms (Table 1). In such pigment mutants, light restores the capability for chlorophyll formation, in most cases up to the level observed in wild-type cells. Transferred back to the dark, chlorophyll biosynthesis ceases. Eventually cells become achlorophyllous by thinning out the remaining chlorophyll during growth. There is only one exception for which a chlorophyll degradation is reported: *Chlorella protothecoides.*[22] This alga bleaches in the presence of glucose and in the absence of nitrogen, and decomposes the chlorophyll at the same time.

The general strategy pursued by green algae under nonpermissive growth conditions and in pigment mutants seems to be to shut off chlorophyll biosynthesis, not necessary in the dark. A light signal immediately regenerates chlorophyll formation. In all cases studied it

Table 1
CONDITIONS OF GREENING FOR GREEN ALGAE AND THEIR MUTANTS HAVING NO CHLOROPHYLL WHEN GROWING IN THE DARK

Organism	Physiological greening conditions	Light	ALA[a]	Chl[a]	Ref.
Chlorella fusca	Regreening after nitrogen starvation	White; 19,000 lux	+	+	4
Chlorella pyrenoidosa	Regreening after nitrogen starvation	White; 500—20,000 lux	+	+	5
Chlorella vulgaris	Regreening after nitrogen starvation	White; 500 lux	+	+	6
Chlamydomonas reinhardii (mutant T4)	Greening by transfer; $37 \rightarrow$ 25°C	(± white)		+	7
Chlorella pyrenoidosa (mutant TS)	Greening by transfer; $31 \rightarrow$ 25°C	(± white)		+	8
Chlorella protothecoides	Regreening after glucose bleaching	Blue; 5×10^{-11} E cm^{-2} sec^{-1}	+	+	9
Chlamydomonas reinhardii (mutant Y-1)	Greening under illumination	White fluorescent; 1016 lux		+	10
Chlorella fusca (mutant G 10)	Greening under illumination	White; 2,500 lux		+	11
Chlorella pyrenoidosa (g-2)	Greening under illumination	White; 10,000 lux		+	12
Chlorella vulgaris (mutant 69)	Greening under illumination	White; 2,500 lux		+	13
Chlorella vulgaris (mutant y_1)	Greening under illumination	White; 8000 lux	+	+	14
Chlorella vulgaris (mutant 10-y)	Greening under illumination	White; 3 W · m^{-2}		+	15
Chlorella vulgaris (mutant 5/520)	Greening under illumination	White; 5,000 lux		+	16
Chlorella regularis (mutant YG-6)	Greening under illumination	Tungsten lamp; 40 W · m^{-2}		+	17
Scenedesmus obliquus (mutant C-2A')	Greening under illumination	Blue; 2.5×10^{-1} E cm^{-2} sec^{-1}	+	+	18
Scenedesmus obliquus (mutant C-2A')	Greening under illumination	White; 10,000 lux	+	+	19
Scenedesmus obliquus (mutant C-6D)	Greening under illumination	White; 10 W · m^{-2}		+	20
Euglena gracilis[b]	Greening under illumination	sodium-vapor lamp; 4700 lux	+	+	21

[a]   Investigations on formation of 5-aminolevulinic acid (ALA) and chlorophyll (Chl) are marked with a +.

[b]   Since its pigment composition is the same as in green algae, *Euglena* is listed here in addition.

is predominantly blue light which is responsible for excitation of the photoreceptor involved.

B. 5-Aminolevulinic Acid Formation

Algae greening only in the light appear phenotypically like angiosperms. They are yellow in the dark and form chlorophyll when transferred into light. One should expect that the limiting light-dependent step would be the same as in flowering plants, i.e., the photoreduction of protochlorophyllide to chlorophyllide. However, only in one case was it reported that protochlorophyllide reduction is the light-limiting step in regreening.[17] In most cases it is not this last step of chlorophyll biosynthesis which becomes light dependent, but the early step of 5-aminolevulinic acid (ALA) formation.

Such a light dependence of ALA formation has been shown for several green algae (Table 1) and in those cases in which monochromatic irradiation was applied, it became clear that

Table 2
**WAVELENGTH PEAK FROM ACTION SPECTRA AND SATURATING
INTENSITIES OF BLUE LIGHT-DEPENDENT RESPIRATION
ENHANCEMENT IN GREEN ALGAE**

| | Action peaks (nm) | | Saturating intensity | |
Organism	Lower	Major	(W · m²)	Ref.
Chlorella pyrenoidosa (211-8b)	375	462	5.1 (flashes)	26
C. pyrenoidosa (211-8b)	370	455	0.5	27
C. vulgaris (211-11h)	382	456	1.0	28
C. vulgaris (211-11h, mutant 125)	382	456	0.5	29
C. vulgaris (211-11h, mutant 20)	370	460	0.5	30
Scenedesmus obliquus (D₃, mutant C-2A′)	390	450	0.4	31
S. obliquus (D₃)	?	458	2.5	32
S. obliquus (D₃, mutant C-6E)	370	468	0.6	See Fig. 1

blue is the most effective wavelength.[9,19] In addition to blue light-controlled ALA formation as the main effect, reactions with minor efficiency can become additionally light dependent (see below).

The action of blue light on ALA formation is manifold. Theoretically it can act via (1) the activation of an existing enzyme, (2) the formation of the ALA synthesizing enzyme(s), or (3) the provision of necessary precursors.

No evidence has been found for the activation of a preexisting inactive enzyme during greening (1). The presence of cycloheximide, the potent inhibitor of protein formation, prevented any induction of ALA and chlorophyll synthesis.[18] When light was turned off for cells with active ALA synthesizing enzymes, ALA formation ceased and enzymes became inactivated. Only a partial reactivation could be induced by light during in vitro tests.[20] This partial reactivation was not inhibited by cycloheximide.

A light-induced form of ALA synthesizing enzyme(s) was demonstrated by in vitro tests.[23] Since this process was inhibited by cycloheximide, these experiments suggest a light-induced *de novo* formation of the ALA-synthesizing enzyme(s) (2). The enzyme formation was accompanied by an increase in soluble proteins.[23] In several cases investigated, a blue light-dependent formation of precursors was the necessary presumption for chlorophyll biosynthesis (3). It is well documented that storage starch is metabolized and respiration is enhanced by blue light.[24,25]

The action spectra measured for respiration enhancement in various organisms are remarkably similar in the position of the peaks and the saturating intensities (Table 2). This blue light reaction can be completely substituted by the addition of glucose to the nutrient medium of the resting cells.

That blue light-induced respiration is the presumption for subsequent greening in pigment mutant C-2A′ of *Scenedesmus* was demonstrated both by action spectroscopy[33,34] and by the fact that greening in this mutant requires the presence of oxygen.[23]

The respiration enhancement and chlorophyll formation in the pigment mutant have similar action spectra (Figure 1). The deviation in the action spectra of respiration enhancement and chlorophyll biosynthesis is probably due to experimental conditions. This interpretation is supported by the action spectrum for respiration enhancement of another pigment mutant (C-6E) of the same strain of *Scenedesmus*. The deviation between the peaks of the action spectra for ALA and chlorophyll formation are well within the limits of the other spectra reported (cf. Table 2). Thus, we might conclude that under these conditions the respiration enhancement is the rate-limiting reaction for chlorophyll biosynthesis.

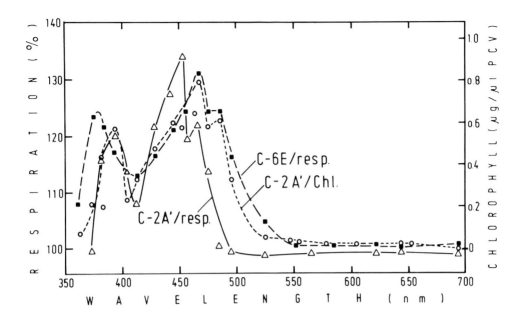

FIGURE 1.   Wavelength-dependent respiration enhancement and chlorophyll formation in pigment mutants of *Scenedesmus obliquus*. Respiration enhancement was measured after 3 hr, chlorophyll content after 15 hr of monochromatic irradiation of 1.75×10^{-10} cm^{-2} sec^{-1}. Cells were harvested for the experiment from the resting phase after 6 days of heterotrophic growth. Pigment mutant C-6E has no detectable caroteinoids and only Chl a and Chl RCI (cf. Wellburn et al.[35]).

Considering the action spectra for ALA formation in *Scenedesmus* mutant C-2A′ published over the years,[18,31,36] one recognizes three facts: (1) the most efficient formation of ALA is always in the blue; (2) there are effects in the red and green region; and (3) spectra show considerable variation. The variation might be explained by the different ages of the cells. According to this parameter, the availability of precursors might vary.

A new action spectrum for ALA formation has therefore been carried out with cells just during transition from the logarithmic to the stationary growth phase (3 days old); in this stage, cells are most receptive for induction of ALA formation (Figure 2). The action spectrum of ALA formation is compared with that of chlorophyll biosynthesis. In this case cells were provided with newly added glucose in order to substitute the blue light induction of respiration and to circumvent this rate-limiting step (Figure 2). In addition to the peaks in the blue, these action spectra for ALA biosynthesis reflect some activity in the red and green regions of the spectrum. The peaks in the red region coincide with those reported for protochlorophyllides, and the green one might be attributed to a hemoprotein or protoheme (see Senger et al.[23] for discussion) as photoreceptor. For chlorophyll formation, the peak in the green is missing and the peaks in the red are identical with those of protochlorophyllide absorption. In order to discuss these action spectra, ALA and chlorophyll biosynthesis must be considered in more detail.

ALA, the first specific intermediate of porphyrin biosynthesis, just forms a transient pool and can only be detected when it is accumulated by blocking the next metabolic step to porphobilinogen (PBG) by the competitive inhibitor levulinic acid (LA). So far the attempt to detect ALA in dark-grown mutant cells with this technique has failed.

Biosynthesis of ALA occurs in algae[38] and higher plants[39] via two separate pathways (Figure 3). According to recent investigations,[36,40,41] the porphyrin of the chlorophyll is only formed via the C-5 pathway and is located in the chloroplast.[42] Thus one has to conclude that the Shemin pathway is located outside the chloroplast and (at least) produces the extraplastidal heme derivatives. In dark-grown mutant cells ALA has never been accumulated

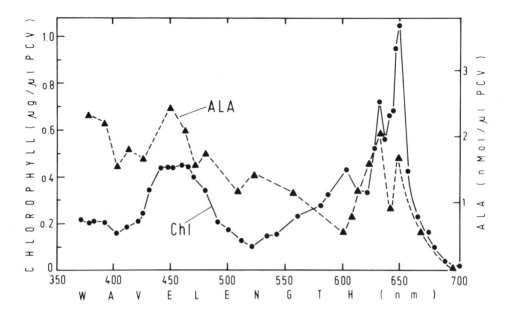

FIGURE 2.   Wavelength-dependent formation of ALA and chlorophyll in pigment mutant C-2A' of *Scenedesmus obliquus*. ALA and chlorophyll content were measured after 16 hr of monochromatic irradiation of 1.75×10^{-10} cm^{-2} sec^{-1}. Cells were used for light-dependent respiration and chlorophyll induction after 3 days of heterotrophic growth. Glucose (0.5% final concentration) and levulinic acid (LA) (10 mM) were added to the samples for induction of chlorophyll and ALA biosynthesis respectively, prior to the start of monochromatic irradiation.

in the presence of LA to a detectable amount. But, since cells grow in the dark, and synthesize cytochrome and protochlorophyllide, one has to conclude that both pathways are operating with very low capacity in the dark. Protochlorophyllide is not transformed and performs a feedback inhibition on the ALA-synthesizing enzyme.[43,44] When exposed to monochromatic light, the protochlorophyllide is transformed to chlorophyllide and the inhibition is taken away from the ALA-synthesizing enzyme. The addition of LA will block the formation of new protochlorophyllide and no feedback inhibition will block the accumulation of ALA. This explains the peaks in the red region of the action spectra for ALA formation and for chlorophyll formation in the presence of glucose. In analogy one has to assume that some compound of the Shemin pathway performs a negative feedback on an ALA-synthesizing enzyme and that this compound is photoconverted by green light and thus allows ALA accumulation in the presence of LA. The blue part of the action spectra is due to the specific effect of this light quality on the *de novo* synthesis of the ALA-synthesizing enzymes.[23]

Summarizing, one may say that several light-dependent processes contribute to porphyrin biosynthesis: photosynthesis (blue and red light), respiration enhancement (blue light), phototransformation of protochlorophyllide (blue and red light), effects on hemoproteins (?; green light). However, only the light-dependent formation of ALA-synthesizing enzymes remains a true and direct blue light effect on prophyrin biosynthesis.

III. CHLOROPHYLL BIOSYNTHESIS IN HIGHER PLANTS

A. Greening Seedlings

Chlorophyll biosynthesis in greening seedlings is controlled at two levels: ALA formation and protochlorophyllide/chlorophyllide reduction (cf. Kasemir[45]). The photoreceptor systems involved are phytochrome (red/far-red) and protochlorophyllide (blue/red). The photoconversion of protochlorophyllide to chlorophyllide removes the feedback inhibition of the ALA-

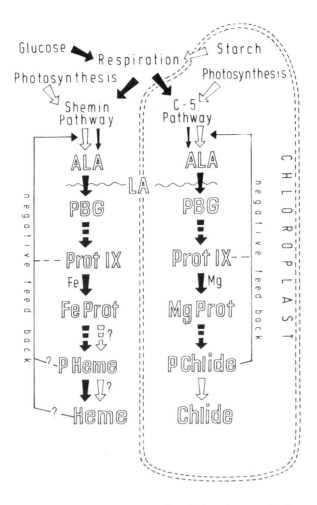

FIGURE 3. Schematic diagram of porphyrin pathways and their regulation. Dark arrows indicate reactions independent of light and the white arrows light-dependent reactions. Known and presumed (?) negative feed backs on ALA-synthesizing enzymes are indicated. ALA = 5 aminolevulinic acid, PBG = porphobilinogene, PROT = protoporphyrine, P Chlide = proto-chlorophyllide, LA = levulinic acid.

synthesizing enzyme(s).[46] The phytochrome system potentiates the ALA formation through activation of the involved enzyme(s).[47,48] A similar potentiation of chlorophyll biosynthesis mediated by a blue light receptor is reported for *Euglena*.[49] For barley seedlings there is a singular report that blue light-grown seedlings have more specific activity for ALA-synthesizing enzymes than seedlings grown under red light.[59]

An increase in the activity of ALA dehydratase, catalyzing the condensation of ALA to porphobilinogen, by long-term far-red illumination[50] does not seem to be a direct light effect.[45] In the green alga *Scenedesmus,* an increase in ALA dehydratase activity was also interpreted as an indirect effect via a substrate activation.[23]

B. Tissue Cultures

Unlike intact plant seedlings, greening tissue cultures demonstrate specific photocontrol of chlorophyll biosynthesis by the blue light system. Chlorophyll biosynthesis in tissue cultures of tobacco,[51-54] potato tubers,[55] and wheat roots[56] proceeds only under blue light.

Likewise, in greening algal cultures the provision of carbon sources seems to be important

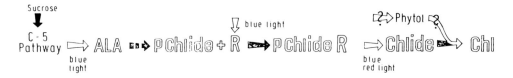

FIGURE 4.   Schematic diagram of chlorophyll biosynthesis in greening tissue cultures and cell suspensions of angiosperms. Dark arrows indicate reactions independent of light and the white arrows light-dependent reactions. Abbreviations as in Figure 3; R = reductant for protochlorophyllide.

for greening of callus cultures of higher plants. In tobacco tissue cultures, respiration and ALA formation increase with the concentration of added sucrose.[52] On the other hand, irradiation with blue light in the same system enhances only ALA formation and not respiration. In other experiments with a tobacco tissue culture of a different species, sucrose was proven to be a necessary prerequisite of greening, but its concentration had no influence on the amount of chlorophyll formed per unit of fresh weight.[54] Glucose can substitute sucrose only to a certain extent. It is taken up under blue light quickly and metabolized immediately. Only under high concentrations will an "overflow" of energy and precursors from general metabolism to chlorophyll biosynthesis take place and enhance its formation.[54]

A direct effect of blue light on ALA accumulation under LA was found in the tissue cultures of tobacco.[52,57] It is intensity dependent, saturates at about 0.6 to 0.8 W m^{-2}. Red light does not stimulate ALA formation. The induction of ALA biosynthesis by blue light is very slow, which seems to be an indication of a *de novo* synthesis of ALA-synthesizing enzymes.[52]

An additional effect of blue light on chlorophyll biosynthesis in tissue cultures is discussed by Kamiya et al.[52] From the consideration of different intensities of blue and red light during subsequent steps of protochlorophyllide transformation, they conclude that blue light of low intensity stimulates the dark formation of the reductant of protochlorophyllide reduction.[58]

Although no action spectra for chlorophyll biosynthesis have been performed, there is enough evidence from the irradiation with single red and blue wavelengths under various light regimes and from fluorescence studies to conclude that protochlorophyllide reduction is light dependent in cells of tissue cultures as it is in seedlings of higher plants. Emerging from the investigations of tissue cultures of higher plants seems to be the scheme proposed by Kamiya et al.,[52] which is simplified in Figure 4.

IV. SUMMARY

When comparing the light regulation of greening algae and tissue cultures of higher plants, similarities become obvious. Direct influence of blue light on ALA-synthesizing enzymes and the feedback control of its activity by photoconversion of protochlorophyllide to chlorophyllide is the common regulatory mechanism. Prerequisite for both greening systems is the provision of energy and precursors. These can be produced either from externally added sugars in the dark or in blue-light-stimulated reactions.

It has to be emphasized that these light regulations are not or are only partially seen in the wild type of the algae and in the intact higher plants. These are potential regulatory mechanisms of chlorophyll biosynthesis which are expressed only under certain circumstances. The similarity in the two greening systems suggests that it is an old regulatory evolutionary mechanism. In this Chapter chlorophyll biosynthesis has only been considered in toto. When branching pathways to different chlorophylls like Chl b, Chl RC I, or divinyl and monovinyl derivatives are examined, the regulatory mechanism might require some modifications.

REFERENCES

1. **Steffens, D., Blos, I., Schoch, S., and Rüdiger, W.,** Lichtabhängigkeit der Phytoakkumulation. Ein Beitrag zur Frage der Chlorophyll-Biosynthese, *Planta,* 130, 151, 1976.
2. **Benz, J., Haser, A., and Rüdiger, W.,** Changes in the endogenous pools of tetraprenyl diphosphates in etiolated oat seedlings after irradiation, *Z. Pflanzenphysiol.,* 111, 349, 1983.
3. **Kasemir, H. and Prehm, G.,** Control of chlorophyll synthesis by phytochrome. III. Does phytochrome regulate the chlorophyllide esterification in mustard seedlings?, *Planta,* 132, 291, 1976.
4. **Porra, R. and Grimme, H.,** Chlorophyll synthesis and intracellular fluctuations of δ-aminolevulinate formation during the regreening of nitrogen-deficient *Chlorella fusca, Arch. Biochem. Biophys.,* 164, 312, 1974.
5. **Meisch, H. U. and Bellmann, I.,** Light dependence of vanadium induced formation of chlorophyll and δ-aminolevulinic acid in *Chlorella., Z. Pflanzenphysiol.,* 96, 143, 1980.
6. **Meller, E. and Harel, E.,** The pathway of 5-aminolevulinic acid synthesis in *Chlorella vulgaris* and in *Fremyella diplosiphon,* in *Chloroplast Development,* Akoyunoglou, G. and Argyroudi-Akoyunoglou, J. H., Eds., Elsevier/North-Holland, Biomedical Press, Amsterdam, 1978, 51.
7. **Kretzer, F., Ohad, I., and Bennoun, P.,** Ontogeny, insertion and activation of two thylakoid peptides required for photosystem II activity in the nuclear temperature sensitive T_4 mutant of *Chlamydomonas reinhardi,* in *Genetics and Biogenesis of Chloroplasts and Mitochondria,* Bücher, Th., Neupert, W., Sebald, W., and Werner, S., Eds., Elsevier/North-Holland Biomedical Press, Amsterdam, 1976, 25.
8. **Lavintman, N., Galling, G., and Ohad, I.,** Repair of photosynthetic activity in thylakoids formed at the non-permissive temperature in A *TS* mutant of *Chlorella,* in *Chloroplast Development,* Vol. 2, Akoyunoglou, G. and Argyroudi-Akoyunoglou, J. H., Eds., Elsevier/North-Holland, Biomedical Press, Amsterdam, 1978, 875.
9. **Oh-hama, T. and Senger, H.,** Spectral effectiveness in chlorophyll and 5-aminolevulinic acid formation during regreening of glucose-bleached cells of *Chlorella protothecoides, Plant Cell Physiol.,* 19, 1295, 1978.
10. **Ohad, I., Siekevitz, P., and Palade, G. E.,** Biogenesis of chloroplast membranes, *J. Cell Biol.,* 35, 521, 1967.
11. **Bauer, K. and Wild, A.,** Die Wirkung von Blaulicht auf den photosynthetischen Elektronentransport bei gelbgrünen Mutanten von *Chlorella fusca, Z. Pflanzenphysiol.,* 80, 443, 1976.
12. **Galling, G.,** Development of thylakoids and photosynthetic activity in thermosensitive and light-dependent mutants of *Chlorella pyrenoidosa Photosynthesis,* Vol. 5, *Chloroplast Development* Akoyunogloū, G., Balaban International Science Services, Philadelphia, 1981, 465.
13. **Wild, A. and Fuldner, K.-H.,** The concentration of cytochrome f and P 700 in chlorophyll-deficient mutants of *Chlorella fusca, Planta,* 136, 281, 1977.
14. **Beale, S. I.,** Studies on the biosynthesis and metabolism of δ-aminolevulinic acid in *Chlorella, Plant Physiol.,* 48, 316, 1971.
15. **Herron, H. A. and Mauzerall, D.,** The development of photosynthesis in a greening mutant of *Chlorella* and an analysis of the light saturating curve, *Plant Physiol.,* 50, 141, 1972.
16. **Dubertret, G. and Joliot, P.,** Structure and organization of system II photosynthetic units during the greening of a dark-grown *Chlorella* mutant, 357, 399, 1974.
17. **Shioi, Y. and Sasa, T.,** Chlorophyll formation in the YG-6 mutant of *Chlorella regularis:* accumulation of protochlorophyllide and protochlorophyll esterified with geranylgeraniol, *Plant Cell Physiol.,* 25, 131, 1984.
18. **Oh-hama, T. and Senger, H.,** The development of structure and function in chloroplasts of greening mutants of *Scenedesmus.* III. Biosynthesis of δ-aminolevulinic acid, *Plant Cell Physiol.,* 16, 395, 1975.
19. **Klein, O. and Senger, H.,** Biosynthetic pathways to δ-aminolevulinic acid induced by blue light in the pigment mutant C-2A' of *Scenedesmus obliquus, Photochem. Photobiol.,* 27, 203, 1978.
20. **Kah, A., Dörnemann, D., Rühl, D., and Senger, H.,** The influence of light and levulinic acid on the regulation of enzymes for ALA-biosynthesis in two pigment mutants of *Scenedesmus obliquus,* in *Photosynthesis,* Vol. 5, *Chloroplast Development,* Akoyunoglou, G., Ed., Balaban International Science Services, Philadelphia, 1981, 137.
21. **Salvador, G. F., Beney, G., and Nigon, V.,** Control of δ-aminolevulinic acid synthesis during greening of dark grown *Euglena gracilis, Plant Sci. Lett.,* 6, 197, 1976.
22. **Oshio, Y. and Hase, E.,** Studies on red pigments excreted by cells of *Chlorella protothecoides* during the process of bleaching induced by glucose or acetate. II. Mode of formation of the red pigments, *Plant Cell Physiol.,* 10, 51, 1969.
23. **Senger, H., Klein, D., Dörnemann, D., and Porra, R. J.,** The action of blue light on 5-aminolaevulinic acid formation, in *The Blue Light Syndrome,* Senger, H., Ed., Springer-Verlag, Berlin, 1980, 541.
24. **Kowallik, W.,** Blue light effects on respiration, *Annu. Rev. Plant Physiol.,* 33, 51, 1982.

25. **Kowallik, W.,** Blue light effects on carbohydrate and Protein metabolism, in *Blue Light Responses: Phenomena and Occurrence in Plants and Microorganisms,* Vol. 1, 1987, chap. 2.

26. **Picket, J. M. and French, C. S.,** The action spectrum for blue-light-stimulated oxygen uptake in *Chlorella, Proc. Natl. Acad. Sci. U.S.A.,* 57, 1587, 1967.

27. **Kowallik, W. and Schätzle, S.,** Enhancement of carbohydrate degradation by blue light, in *The Blue Light Syndrome,* Senger, H., Ed., Springer Verlag, Berlin, 1980, 344.

28. **Miyachi, S., Tsusuki, M., and Miachi, S.,** Effects of blue light on carbon metabolism, in *The Blue Light Syndrome,* Senger, H., Ed., Springer Verlag, Berlin, 1980, 321.

29. **Kamiya, A. and Miyachi, S.,** Effects of blue light on respiration and carbon dioxide fixation in colorless *Chlorella* mutant cells, *Plant Cell Physiol.,* 15, 927, 1974.

30. **Kowallik, W. and Gaffron, H.,** Respiration induced by blue light, *Planta,* 69, 92, 1966.

31. **Brinkmann, G. and Senger, H.,** Light-dependent formation of thylakoid membranes during the development of the photosynthetic apparatus in pigment mutant C-2A′ of *Scenedesmus obliquus,* in *Photosynthesis,* Vol. 5, *Chloroplast Development,* Akoyunoglou, G., Ed., Elsevier/North-Holland Biomedical Press, Amsterdam, 1978, 201.

32. **Kulandaivelu, G. and Sarojini, G.,** Blue light induced enhancement in activity of certain enzymes in heterotrophically grown cultures of *Scenedesmus obliquus,* in *The Blue Light Syndrome,* Senger, H., Ed., Springer-Verlag, Berlin, 1980, 372.

33. **Senger, H. and Bishop, N. I.,** The development of structure and function in chloroplasts of greening mutants of *Scenedesmus.* I. Formation of chlorophyll, *Plant Cell Physiol.,* 13, 633, 1972.

34. **Brinkmann, G. and Senger, H.,** Blue light regulation of chloroplast development in *Scenedesmus* mutant C-2A′, in *The Blue Light Syndrome,* Senger, H., Ed., Springer-Verlag, Berlin, 1980, 526.

35. **Wellburn, F. A. M., Wellburn, A. R., and Senger, H.,** Changes in ultrastructure and photosynthetic capacity within *Scenedesmus obliquus* mutants C-2A′, C-6D, and C-6E on transfer from dark grown to illuminated conditions, *Protoplasma,* 103, 35, 1980.

36. **Senger, H. and Rühl, D.,** Mode of coaction of photosensing pigments, in *Photosynthesis Research,* Sybesma, C., Ed., Martinus Nijhoff/Dr. W. Junk, The Hague, 1984, 795.

37. **Schopfer, P. and Siegelmann, H. W.,** Purification of protochlorophyllide holochrome, *Plant Physiol.,* 43, 990, 1968.

38. **Klein, O. and Senger, H.,** Biosynthetic pathways to δ-aminolevulinic acid induced by blue light in the pigment mutant C-2A′ of *Scenedesmus obliquus, Photochem. Photobiol.,* 27, 203, 1978.

39. **Porra, R. J., Klein, O., and Wright, P. E.,** The proof of ^{13}C-NMR spectroscopy of the predominance of the C_5-pathway over the shemin pathway in chlorophyll biosynthesis in higher plants and the formation of the methyl ester, *Eur. J. Biochem.,* 130, 509, 1983.

40. **Oh-hama, T., Seto, H., Otake, N., and Miyachi, S.,** ^{13}C-NMR evidence for the pathway of chlorophyll biosynthesis in green algae, *Biochem. Biophys. Res. Commun.,* 105, 647, 1982.

41. **Porra, R. J., Klein, O., and Wright, P. E.,** ^{13}C-NMR-studies of chlorophyll biosynthesis in higher plants: an unequivocal proof of the participation of the C_5-pathway and evidence of a new route for the incorporation of glycine, *Biochem. Int.,* 5, 345, 1982.

42. **Kannangara, C. G. and Gough, S. P.,** Synthesis of δ-aminolevulinic acid and chlorophyll by isolated chloroplasts, *Carlsberg Res. Commun.,* 42, 441, 1977.

43. **Burnham, B. F. and Lascelles, J.,** Control of porphyrin biosynthesis through a negative feed-back mechanism, *Biochem. J.,* 87, 462, 1963.

44. **Beale, S. I.,** δ-Aminolevulinic acids in plants: its biosynthesis, regulation, and role in plastid development, *Annu. Rev. Plant Physiol.,* 29, 95, 1978.

45. **Kasemir, H.,** Light control of chlorophyll accumulation in higher plants, in *Encyclopedia of Plant Physiology, New Series Vol. 16B, Photomorphogenesis,* Shropshire, W., Jr. and Mohr, H., Eds., Springer-Verlag, Berlin, 1983, 662.

46. **Ford, M. J. and Kasemir, H.,** Correlation between 5-aminolaevulinate accumulation and protochlorophyll photoconversion, *Planta,* 150, 206, 1980.

47. **Klein, S., Katz, F., and Neeman, E.,** Induction of δ-aminolevulinic acid formation in etiolated maize leaves controlled by two light systems, *Plant Physiol.,* 60, 335, 1977.

48. **Kasemir, H. and Mohr, H.,** The involvement of phytochrome in controlling chlorophyll and 5-aminolevulinate formation in a gymnosperm seedling *(Pinus silvestris), Planta,* 152, 369, 1981.

49. **Schiff, J. A.,** Blue light and the photocontrol of photocontrol of chloroplast development in *Euglena,* in *The Blue Light Syndrome,* Senger, H., Ed., Springer-Verlag, Berlin, 1980, 495.

50. **Kasemir, H. and Masoner, M.,** Control of chlorophyll synthesis by phytochrome. II. The effect of phytochrome on aminolevulinate dehydratase in mustard seedlings, *Planta,* 126, 119, 1975.

51. **Bergmann, L. and Berger, Ch.,** Farblicht und Plastidendifferenzierung in Zellkulturen von *Nicotiana tabacum* "Sumsun", *Planta,* 69, 58, 1984.

52. **Kamiya, A., Ikegami, I., and Hase, E.,** Effects of blue light on the formation of 5-aminolevulinic acid and chlorophyll in cultured tobacco cells, in *Blue Light Effects in Biological Systems,* Senger, H., Ed., Springer-Verlag, Berlin, 1984, 335.

53. **Richter, G., Reihl, W., Wietoska, B., and Beckman, J.,** Blue light induced development of thylakoid membranes in isolated seedling roots and cultured plant cells, in *The Blue Light Syndrome,* Senger, H., Ed., Springer-Verlag, Berlin, 1980, 465.

54. **Richter, G., Hundrieser, J., Gross, M., Schultz, S., Bottländer, K., and Schneider, Ch.,** Blue light effects in cell cultures, in *Blue Light Effects in Biological Systems,* Senger, H., Ed., Springer-Verlag, Berlin, 1984, 387.

55. **Berger, Ch. and Bergmann, K.,** Farblicht und Plastidendifferenzierung im Speichergewebe von *Solanum tuberosum L., Z. Pflanzenphysiol.,* 56, 439, 1967.

56. **Björn, L. O.,** Chlorophyll formation in excised wheat roots, *Physiol. Plant,* 18, 1130, 1965.

57. **Kamiya, A., Ikegami, I., and Hase, E.,** Effects of light on chlorophyll formation in cultured tobacco cells. II. Blue light effect on 5-aminolevulinic acid formation, *Plant Cell Physiol.,* 24, 779, 1983.

58. **Kamiya, A., Ikegami, I., and Hase, E.,** Effects of light on chlorophyll formation in cultured tobacco cells. I. Chlorophyll accumulation and phototransformation of protochlorophyll (ide) in callus cells under blue and red light, *Plant Cell Physiol.,* 22, 1385, 1981.

59. **Girnth, C., Gough, S. P., and Kannangara, C. G.,** Induction of Δ-aminolevulinate synthesizing activity by light, in *Photoreceptors and Plant Development,* De Greef, J., Ed., Antwerpen University Press, Antwerp, 1980, 261.

Chapter 8

BLUE LIGHT EFFECTS IN ENDOGENOUS RHYTHMS

Rainer Schmid

TABLE OF CONTENTS

I. INTRODUCTION

Rhythms may, in a broad sense, be defined as repetitive temporal patterns. In biological systems a rhythm either can be provoked by external signals or it may be an intrinsic or "endogenous" property of the organism. Endogenous rhythms can easily be detected and differentiated from induced ones since they persist under constant environmental conditions.

Endogenous periodic oscillations include a variety of phenomena such as annual, lunar, tidal, and diel rhythms which represent adaptations to geophysical periodicities or the cell cycle. Other endogenous rhythms cannot be simply described as quantitatively changing parameters as, for instance, repetitive developmental steps like branching of some algae or changes in the direction of the mitotic apparatus in the apical cells of mosses or ferns. Consequently, endogenous cycles are features common to all living systems.

In those cases where endogenous rhythms represent adaptations to geophysical cycles, the organisms use them to be prepared at any time to changes in the environment in order to behave appropriately. Furthermore, they can be used for day-time and day-length measurements (temporal orientations) which, in addition, may serve for the orientation in space in insects and birds (see References 1 and 2).

Some of the cyclicly changing environmental parameters are light conditions, such as light intensity, length of the light period, and variations in moonlight. The organisms, when sensitive, can use these cycles to achieve the best coincidence of the endogenous activity rhythms with the geophysical environment. Light, therefore, has a number of distinct effects on endogenous rhythmicity, which are the main subject of this article. In accordance with the aims of the complete volume, this author will summarize only the effects of blue light in endogenous rhythms. The review will be confined only to those effects which are mediated exclusively by light of wavelengths shorter than 520 nm through the action of blue light receptors.[3] Most of the information available deals with blue light effects in circadian rhythms. Thus this chapter will contribute mainly to this topic.

II. CIRCADIAN RHYTHMS

Among endogenous rhythms, the circadian cycles play a prominent role because they are adaptations to the diel light-dark alternations. They are thought to be driven by an endogenous basis oscillation, called "circadian clock", which is regarded to be a property common to all eukaryotic organisms.[1] Several criteria serve to discriminate circadian oscillations from other cycles with periods of similar range.

- Circadian rhythms persist in a constant environment. Under such conditions they are free-running, i.e., the period (time needed to run through one complete cycle) of the rhythm is different from 24 hr. The period of circadian rhythms usually does not fall below 20 or exceed 28 hr. Therefore, being in the range of 1 day, those rhythms were called "circadian" (Latin: *circa* = about, *dies* = day). Corresponding to the geophysical day, the circadian period is, for practical reasons, divided into 24 "circadian hours". An inducing or synchronizing event is defined to set the clock to 0 C.T. (= circadian time of zero).

- External signals, like temperature changes and/or changes in light, can synchronize the rhythm. An oscillation which is out of pace responds to such a stimulus by a phase shift. This is expressed by an insertion of one or a very few either shortened or lengthened periods. These shifts are called phase advances or delays, respectively. Whether the rhythm is reset by an advance or a delay depends on the phase (status of the oscillation within one period) at which the stimulus is applied. Plots of the amount of phase shift, either positive or negative, relative to the previous rhythm vs. the phase at which the stimulus is given are named "phase response curves". Phase responses are thought to indicate that the clock is directly affected. By repeated stimuli the rhythm can be "entrained" also to periods other than that of the natural 24-hr-cycle, but these may not be shorter or longer than the definite limit cycles which are properties of the organism. When these limit cycles are exceeded perturbations are observed or the rhythm begins to free run.

● Within the range of physiological temperatures the clock is temperature compensated. The oscillation is neither accelerated nor retarded and the Q_{10} value is about one.

These criteria have been frequently summarized. For a more detailed information on circadian rhythms see these reviews.[1,2,4-7]

A. Effects of Blue Light

The following summarizes blue light effects on circadian oscillations. Photoreceptors other than those for blue light may influence the rhythms as well. This is the subject of a thorough compilation by Ninnemann.[8] The status of our present knowledge on the interaction of photoreceptive sites with the clock has been summarized by Engelmann.[9]

1. Induction of a Rhythm

The blue light-induction of a circadian rhythm has been reported occasionally. However, before presenting examples, one must introduce the general consideration that the term "induction" is used to characterize the observation that an arhythmic behavior upon a stimulus becomes a rhythmic behavior. This, of course, may be due to setting the clock into operation. On the other hand, arhythmic behavior of an organism or a population may be due to asynchrony in preexistent rhythms in the component cells or individuals. In this case the stimulus has merely synchronizing character, which then makes the rhythm obvious.

The rhythm of egg hatching in the moth *Pectinophora gossypiella* can be induced by a single blue-light pulse, demonstrated by an action spectrum.[10] Light of wavelengths between 390 and 520 nm is almost equally effective, but is ineffective above 520 nm. The eggs are highly sensitive and pulses of blue light shorter than 2 min with a fluence of 10^{-7} moles · m^{-2} suffice to initiate the rhythm. The induction is only possible after the 5th day of embryogenesis.[10] This may be taken as an indication of a true induction. Results like these do not exclude, however, that it is merely the senstivity to light which is time dependent.

In the fungus *Sclerotinia fructicola*, light induces a zonation rhythm of conidiophore formation.[11] Action spectra for this effect, measured on the basis of equal fluence rates show maxima in the near ultraviolet (UV) and in the blue. The rhythm damps out within 3 days and its circadian nature has not yet been demonstrated. Whether the light effect is due to true induction or to synchroniziation of the rhythms of individual hyphae is unclear.

Broad-band blue light evokes a photosynthetic rhythm in populations of the siphonaceous green alga *Acetabularia mediterranea* after previous cultivation in prolonged red light.[12] Although the circadian rhythms were reported to run out in continuous darkness (DD),[13] it could be shown that the circadian organization is not lost in the individual cells under red light conditions and that blue light can entrain the rhythm of chloroplast migration.[14-16] Different circadian rhythms in *Acetabularia* have been shown to remain in their relative phase position, although the circadian period varied greatly.[17] Therefore, it is likely that the "induction" of the photosynthetic rhythm is due to synchronization within the population.

Induction of circadian rhythmicity may not only be caused by switching on the irradiation but also by switching off events. The latter is the case in *Drosophila* where the transition from continuous light to DD can induce the rhythm of pupal eclosion.[18] A switch from bright light to dim light is effective as well.[19] The rhythm of egg release in *Laminaria* gametophytes can also be initiated by switching off the light.[20] The light effect in these cases is due to an inhibition of the rhythm by light rather than an induction by darkness (see below).

2. Synchronization

Synchronization of circadian rhythms is brought about by phase shifting. In almost all cases where blue light has been shown to play a role in circadian cycles, phase resetting is one of the effects observed. As far as it has been investigated, blue light produces the typical phase-response curves (exception *Coleus;* see below).

A detailed action spectrum for phase delays has been measured for *Drosophila pseudoobscura.*[18] It demonstrates clearly the involvement of a blue light receptor. The action spectra for advancing and delaying the rhythm in this fly are identical.[21] While the delay response is completed within one cycle, the advance response requires several transient cycles.[18,21] Chandrashekaran and Engelmann report that the threshold for the sensitivity to blue light is 10 times higher for advancing the rhythm at 19 C.T. than for delaying it at 18 C.T.[22] They explain this by differences in the responsiveness of the oscillating system rather than by changes in the photoreceptor. Ninnemann,[8] however, suggests two different photoreception processes.

The rhythms of egg hatching, of adult emergence from pupae, and of egg-laying activity in *Pectinophora* are very similar in their light phase-response curves, although showing very different phase relationships to an entraining external light-dark cycle.[23,24] However, they differ in their free-running period and their sensitivity to red light (600 nm).

The amount of phase shifting as a function of the duration of a triggering blue light pulse of constant fluence rate has been shown to have two different saturation levels in the *Neurospora crassa timex* mutant.[25] This has been taken as an indication for two light reactions participating in the response. However, biphasic fluence response curves obtained for blue light-dependent carotenogenesis in *Neurospora* have been interpreted as temporal insensitivity of the photoreception system.[26] The action spectrum for clock resetting shows the usual characteristics of blue light receptors.[27,28]

In a mutant of *Trichoderma harzianum,* which undergoes a circadian banding pattern of conidiation, a pulse of blue light produces two effects: an extra band of conidia and phase shifting of the circadian oscillation with typical phase-response behavior.[29]

When *Acetabularia mediterranea* is grown in continuous red light, the addition of blue light induces phase shifts (Figure 1). No phase shifting occurs when red light is switched on during constant blue light. This proves that only blue light is effective. Advances as well as delays are observed.[15,16]

An unusual result was obtained with *Coleus* (Lamiaceae).[30] Phase shifts in the rhythm of leaf movement were observed with red and blue light. Red light causes advances and blue light delays. The combined effects of both light qualities results in the phase-response curve which is found for white light. No blue light response on shifting was found in the phases of red light responsiveness and vice versa.

3. Period Length

Continuous light can influence the period of free-running oscillations. According to a rule formulated by Aschoff,[31] the period shortens with increasing light intensities in day-active organisms, while in night-active organisms it becomes longer. One of the exceptions to this rule is *Acetabularia.* In the circadian chloroplast migration under red light, not only did the amplitude decrease with the duration of the illumination, but also very short periods, ranging between 20.5 and 22 hr were found.[15,16] Lengthened periods are produced by raising the blue light intensity on a constant red light basis (Figure 1). At high blue fluence rates, they are as long as 29 hr. The new periods are often adapted only after passing through two to four transient cycles. Changes in the circadian period are not caused by varying the red light

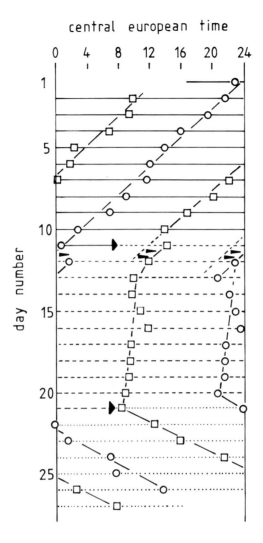

central european time

FIGURE 1.   Phase plot of the chloroplast migration in *Acetabularia mediterranea*.[16] Times of minimum chloroplast density at the cell apex are indicated by circles, and maximum densities by squares. The solid line indicates irradiation with red light only (3.5 W · m⁻²). Broken line: blue light at a fluence rate of 16 W · m⁻² was added. The onset of the blue irradiation induced a delay of 2 hr (arrowheads). Dotted line: the blue light fluence rate was increased to 160 W · m⁻².

fluence rate on a constant blue light background. It can be excluded, therefore, that the effects are mediated through photosynthesis.

The period of the rhythm of leaf movement in entrained *Coleus* plants (LD 12:12, white light) was slightly shortened to 2.5 hr when the plants were moved to DD, green, or far red light. Continuous red light evoked a very short period of 20.5 hr, while blue light keeps the period at 24 hr.[30]

4. Inhibition of the Rhythm

A rather frequent phenomenon is that the circadian rhythm is lost at high light intensities. As in the case of rhythm induction, some theoretical preconsiderations should be kept in mind. First, true inhibition of the clock may be possible. Second, the inhibiting condition may lead to desynchronization in cells or individuals in a population. Finally, the rhythmic activity, which can be taken as the visible expression of the endogenous clock, may be uncoupled from a basis oscillator, although the clock is still running. In most of the cases discussed below these possibilities have not been considered sufficiently.

The *Drosophila* pupae emergence rhythm damps out within 3 days, when they are irradiated with continuous blue light of more than $1 \text{ W} \cdot \text{m}^{-2}$.[19] Below this value the rhythm damps out as well, but the lower the light intensity, the later the rhythm disappears. The light intensity of $0.01 \text{ W} \cdot \text{m}^{-2}$ is below the threshold. No action spectrum has been measured for this effect. Photoreception through the visual system is unlikely.

The circadian conidiation rhythm in the *Neurospora* clock mutants can be suppressed by blue light.[32] Sargent and Briggs have shown an action spectrum for this effect[25] which, again, is typical for blue light receptors.

Gametophytes of the brown macroalga *Laminaria* release eggs only in darkness. In DD this egg release is rhythmic with a period of about 1 day and persists until the 11th day. The rhythm is, therefore, probably circadian. Blue light inhibits the release of eggs and the action spectrum matches those for other blue light effects.[20] Since the release of eggs can be shifted to any point of time by prolonging the irradiation period of the normal LD cycles, the effect of blue light is to inhibit the release rather than to stop the rhythm.

Only a few experiments have been described in which a pulse of light could successfully inhibit a circadian rhythm. A single flash (50 sec at $0.01 \text{ W} \cdot \text{m}^{-2}$) of blue light, given to *Drosophila* pupae, at the phase transition point at 18.6 C.T. (18.6 hr after the entraining transfer from light to DD) leads to an attenuation of the amplitude of the eclosion rhythm.[33] The reciprocity law is valid.[34] In this kind of experiments the blue light pulse apparently evokes a true arrest of the clock.[33] With prolonged exposure to darkness the sensitivity to blue light increases by a factor of 10 within 24 hr.[35] This was interpreted as a recovery of the photoreceptor in continuous darkness from bleaching under the continuous light regime before the start of the experiment.

B. Photoreception and Light Signal Transduction

In all representative blue light responses the shapes of the action spectra point to two different pigment classes as possible photoreceptors, namely, the carotenoids or the flavins (see other chapters, this volume). It is evident that the chemical identity of the receptor as well as the subsequent reactions have to be elucidated for each organism separately. Action spectroscopy as the only tool is insufficient for this purpose. Additional approaches to clarify the photoreceptive mechanism have been applied only to two circadian systems.

1. Drosophila

Eyeless mutants of *Drosophila melanogaster* can still be entrained by light.[36] The light reception system for vision, therefore, is not responsible for light effects in the circadian rhythm. A theoretical consideration renders this unlikely as well: about 1 week would pass to half saturate the receptors for vision at light intensities sufficient to reset or to inhibit the clock.[19] Therefore, the visual system would be insufficient for this purpose. Furthermore, carotenoid derivatives, retinal included, are unlikely to be the photoreceptors, since *D. melanogaster* grown on a carotenoid-depleted diet, lose sensitivity in the visual system by a factor of 1000, while the sensitivity of photoreception in the circadian rhythm is unaffected.[37] A similar experiment with *D. pseudoobscura* resulted in larger phase shifts compared to the controls. Under these conditions of carotenoid deficiency larger light induced absorbance changes (LIACs) were found as well,[38] thus indicating that the photoreceptor may be a flavin.

As mentioned above, the threshold sensitivity of *D. pseudoobscura* pupae for blue light is different by a factor of 10 for advancing or delaying the clock.[22] The phase points in

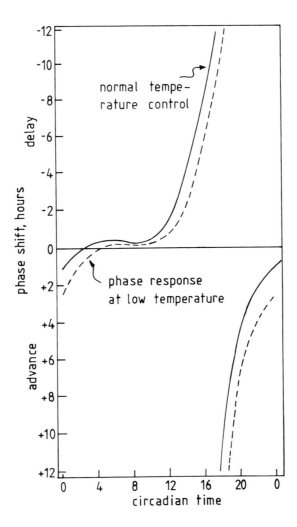

FIGURE 2. Light phase-response curve of pupae eclosion in *Drosophila pseudoobscura*.[40] The solid line represents the light phase-response curve at normal temperature (20°C). The phase response for light during pulses of low temperature (6°C), after correction for the effects of temperature pulses alone, is shown by the dashed line. It is shifted relative to the normal response curve by about 2 hr, a time which corresponds to the duration of the low-temperature pulse.

these experiments at which the fluence response was measured were only 1 hr apart. A tenfold increase in sensitivity within this time seems to be unlikely. Thus, these differences were attributed to altered properties of the circadian clock. Ninnemann,[8] in reviewing these results, objected that either two different photoreceptors or two different primary processes may be involved in advancing or delaying the clock.

Although the clock itself is temperature compensated and the phase-response curves for light remain unchanged in a wide range of physiological temperatures,[39] it appears that there are temperature-sensitive steps between the photoreceptor and the clock. Hamm et al.[40] investigated the phase-shifting effect of light during pulses of low temperature. After correcting for the effects of temperature pulses alone, they found the phase-response curves for light were shifted (Figure 2). This shift could be attributed to the period which had elapsed between light pulse and raising the temperature to the normal level. Thus, like blue light reception in carotenogenesis in fungi,[41] photoreception was possible at low temperatures. However, the information remained stored until the temperature was raised to a physiological level. Furthermore, the experiments indicate that photoreception is not a property of the clock itself.

2. Neurospora crassa *Clock Mutants*

The albino-timex mutant of *Neurospora crassa,* which lacks major carotenoids, is unaltered in sensitivity to blue light. At least a major carotenoid fraction is therefore unlikely to act as the photoreceptor.[25] Some results point to the possible involvement of the flavo-hemomolybdo-enzyme nitrate reductase in the photoreception in *N. crassa.* Blue light stimulates the activity of the small subunit of the enzyme after previous starvation of the mycelia. This stimulation coincides well with light-induced conidiation and a LIAC.[42,43] Furthermore, nitrate reductase can be light stimulated in partially purified preparations and the effect shows the characteristic action spectrum of the blue light effects.[44]

In experiments with three nitrate reductase mutants of *Neurospora,* Paietta and Sargent detected no significant differences in the phase shifting and photoinhibition response as compared to the normal clock mutant.[45] Furthermore, under conditions which inhibit nitrate reductase activity or suppress the formation of the enzyme, the responses of the conidiation rhythm to blue light are normal.[45] It has to be concluded that nitrate reductase plays a secondary role in light reception for the circadian rhythm, if any.

There is strong evidence that the photoreceptor in *Neurospora* is a flavin. In mutants auxotroph for riboflavin a good correlation was found between riboflavin concentration in the culture and the sensitivity to blue light in phase shifting and suppression of the rhythm.[46] Further evidence comes from experiments using riboflavin analogs, which can replace the natural receptor in *Neurospora.* Deazariboflavin and roseoflavin, which have green-shifted absorption spectra, are responsible at 540 nm to phase shifting and photoinhibition, while the untreated mycelia do not respond at this wavelength.[47]

Brain et al.[48] have found LIACs in plasma-membrane preparations of *Neurospora.* LIACs are due to the reduction of a b-type cytochrome by a flavin.[49] They can be sensitized for red light by methylene blue.[50] Since red light in the presence of methylene blue can inhibit and phase shift the circadian rhythm in *Neurospora* (Briggs unpublished, cited in Reference 27) a LIAC might be the primary reaction in the photoreception of the circadian rhythm. This idea is in further agreement with observations on the *N. crassa* "poky" mutant, which is characterized by a reduction of the total cytochrome content (extramitochondrial cytochrome to 16%). The cytochrome deficiency is accompanied by a strong loss of sensitivity of the mutant to blue light with respect to the inhibition of the circadian rhythm.[51] On the other hand, no causal relationship could be found between a LIAC and light-induced phase shifting in the albino band mutant, a relationship which was found, however, for light-induced conidia formation.[52] Thus, different photoreception processes for blue light may act in *Neurospora.* Furthermore, Paietta and Sargent obtained evidence that blue light photoreception in circadian conidiation might be different from that in carotenogenesis.[46,47] This was concluded from their experiments with *Neurospora* riboflavin auxotrophs as well as from studies with riboflavin analogs.

The process which couples the clock to photoreception was found to include heat-labile steps.[53] Plasma-membrane ATPase inhibitors can suppress light-induced phase shifting under certain conditions.[54] It was suggested, therefore, that a flavin-associated photoreceptor could stimulate a proton-translocating plasma-membrane ATPase leading to phase shifts. The effects of the inhibitors, however, are difficult to interpret, and the involvement of the ATPase in the blue light-mediated phase shifting is ambiguous.[55]

At high pH (>7.6) light-induced phase resetting requires the presence of ammonium ions in the culture medium at the time of the irradiation. In contrast, phase shifting at low pH

(5.7) was independent of external ammonium.[56] The effects of ATPase inhibitors on light-induced phase shifting exhibit similar pH dependence.[54] The requirement for ammonium ions is interpreted, therefore, as demonstrating their regulatory role in specific biochemical reactions such as ATP-dependent proton translocations in the plasma membrane.[56]

C. Circadian Rhythms of Blue Light Responses

The preceding sections have described the effects of blue light on circadian rhythms. Occasionally a blue light response itself may vary with a circadian periodicity. In the fungus *Trichoderma harzianum,* changes in the sensitivity of induction of sporulation by blue light follow a circadian rhythm.[29] The "stop response" in the phototactic movement of the dinoflagellate *Gyrodinium dorsum* is mediated by blue light. The occurrence of the response requires a brief pre-irradiation of red light which, in addition, causes a shift in the action spectrum. In this organism, as well, the sensitivity to blue light (threshold) changes in a circadian manner.[57] Blue light pulses enhance the dark respiration rate in *Chlorella fusca*. In this case the magnitude of the response to a saturating blue light irradiation varies, following a circadian rhythm.[58]

D. Adaptive Significance of Light Sensing for Circadian Rhythms; Photoperiodism

Two different functions of the endogenous diurnal oscillations are evident. Both require that the endogenous rhythm be synchronized to the geophysical day-night cycle by sensing light or temperature signals.

In the first function, circadian rhythms serve as a mechanism to "know" the time of the day. Physiological activities change diurnally and the organism can respond in the most favorable way during the diel variations of the environmental conditions.

In the second function, circadian rhythms serve as a mechanism to adapt to the seasons of the year. Variations of the circadian period due to the (blue) light intensity, as for instance in *Acetabularia*,[14,16] may be interpreted as a kind of seasonal adaptation. In summer light intensities are higher than in winter. This is coupled to longer irradiation periods. Thus, the effects of light intensity on the period may help to exploit the prolonged summer light periods through a possible alteration in the shape of the oscillation, although this has not yet been proven experimentally. *Acetabularia* shows circadian oscillations in photosynthetic activity.[59] In addition, the chloroplast density at the apex is higher during the circadian day.[60] Both rhythms allow the alga to utilize a long light period more efficiently for photosynthesis in that the phases of high physiological activity are prolonged by the higher light intensities. This interpretation is in contrast to the implications of Aschoff's rule, which, however, has been worked out primarily for animals.

Another generally accepted and more precise mechanism for seasonal adaptation is observed in photoperiodic phenomena. Here, the circadian system is used to measure the length of the daily light or dark periods, respectively. The general principle is that the organisms go through photophilic and photophobic phases which alternate in a circadian way. Light in the photophilic phase promotes the realization of a photoperiodic phenomena. In the photophobic phase photoperiodic phenomena are suppressed. Whether the photoperiodic event is expressed in short days or in long days depends on the phase positions of the phobic or philic phases relative to the diel day/night cycle. In some cases the radiation effective in photoperiodism has been shown to be blue light.

Fruiting body formation in the fungus *Coprinus congregatus* is a short-day phenomenon.[61] Blue light pulses in the night phase suppress the development of fruiting body primordia. Elevated temperatures support this effect. The brown alga *Scytosiphon lomentaria* produces erect thalli only under short-day conditions.[62] The critical day length is between 12 and 13 hr. Only four short-day cycles are required for the transition from the growth as crusts to the formation of cylindrical thalli. An action spectrum shows clearly the involvement of blue light only in the inhibitory effect of long days on thallus formation. In the long-day plant *Sinapis alba,* flowering can be induced under short-day conditions by short light breaks within the night phase.[63] A crude action spectrum shows that only blue light is effective.

III. OTHER ENDOGENOUS RHYTHMS; CELL CYCLE

Regularly alternating endogenous processes are widespread in biological systems and occur often as a principle in pattern formation for development. The influence of light on such phenomena has frequently been documented. However, it would be beyond the limits of this chapter to follow up all those blue light effects which play a role in endogenous oscillations. I shall, therefore, restrict myself to a very few examples.

The most important noncircadian cycle is the cell cycle. Mitotic activity in protonemata of the fern *Pteris vittata* is suppressed in red light. Blue light releases the cell cycle from an arrest in the G_1 phase and induces synchronous divisions.[64] The effect of a short pulse of blue light can be partially reversed when red light is given immediately after blue light.[65] Higher blue light fluences overcome the inhibiting effect of red light. In the gametophyte of the fern *Adiantum,* the G_1 phase of the cell cycle was shortened by blue light to about 40% as compared to darkness.[66,67] Photoreception for this effect was shown to occur in the nuclear region.[66,67]

In the fungus *Leptosphaeria michotii,* a blue light effect has been found in the stabilization of a noncircadian rhythm.[68]

IV. CONCLUDING REMARKS

Light effects in the rhythms can be discriminated with respect to the mode of irradiation. Light pulses may synchronize or induce the rhythm. Phase shifting appears to be due to a process of quanta counting.[18,25] Continuous irradiation affects the period length or inhibits rhythmicity. The effect of light in this case is dependent on the fluence rate. One should expect that the same photoreceptor is involved, wherever both kinds of light sensing are found in one organism *(Drosophila, Neurospora, Acetabularia).* A photoreceptor pigment for blue light in cooperation with the linked primary processes should be able to serve in both functions. This should be kept in mind when looking for possible receptor candidates. On the other hand, various primary processes may cooperate in the same phenomenon, because advances and delays in the circadian rhythm can be induced by different photoreceptors.[30] They can also differ with regards to the sensitivity to blue light.[22]

It is clear that the photoreceptive site cannot be considered as an integral part of the circadian clock.[8,9,40,53] The mechanism of the clock is still unknown. However, the elucidation of the process of light signal transduction may lead to the identification of properties of the circadian basis oscillator. A number of different models have been proposed,[69-73] only some of them including photoreceptive processes.[66-68] None of the models completely explain all phenomena observed for light action.[9] Furthermore, the question is open whether the mechanism of the underlying oscillator is identical in all organisms.

REFERENCES

1. **Bünning, E.**, *Die physiologische Uhr*, Springer-Verlag, Berlin, 1977.
2. **Hastings, J. W. and Schweiger, H.-G.**, *The Molecular Basis of Circadian Rhythms*, Abakon Verlag, Berlin, 1976.
3. **Senger, H.**, Cryptochrome, some terminological thoughts, in *Blue Light Effects in Biological Systems*, Senger, H., Eds., Springer-Verlag, Berlin, 1984, 72.
4. **Queiroz, O.**, Circadian rhythms and metabolic patterns, *Annu. Rev. Plant Physiol.*, 25, 115, 1974.
5. **Hillman, W. S.**, Biological rhythms and physiological timing, *Annu. Rev. Plant Physiol.*, 27, 159, 1976.
6. **Winfree, A. T.**, *The Geometry of Biological Time*, Springer-Verlag, Berlin, 1979.
7. **Pittendrigh, C. S.**, Circadian organization and the photoperiodic phenomena, in *Biological Clocks in Seasonal Reproductive Cycles*, Follet, B. K. and Follet, D. E., Eds., Wright, Bristol, 1981, 1.
8. **Ninnemann, H.**, Photoreceptors for circadian rhythms, in *Photochem. Photobiol. Rev.*, Vol. 4, Smith, K. C., Eds., Plenum Press, New York, 1979, 207.
9. **Engelmann, W.**, Photoreception and the clock, in *Biological Clocks in Seasonal Reproductive Cycles*, Follet, B. K. and Follet, D. E., Eds., Wright, Bristol, 1981, 37.
10. **Bruce, V. G. and Minis, D. H.**, Circadian clock action spectrum in a photoperiodic moth, *Science*, 163, 583, 1969.
11. **Jerebzoff, S. and Jaques, R.**, Equal quantal spectra for the effect of light on the growth of conidiophores and for the induction of a circadian rhythm of zonation in *Sclerotinia fructicola* (Wint.) Rehm., *Plant Physiol.*, 50, 187, 1972.
12. **Clauss, H.**, Auslösung der circadianen Photosynthese-Rhythmik bei *Acetabularia* durch Blaulicht, *Protoplasma*, 99, 341, 1979.
13. **Vanden Driessche, Th.**, Circadian rhythm in *Acetabularia*: photosynthetic capacity and chloroplast shape, *Exp. Cell Res.*, 42, 18, 1960.
14. **Schmid, R.**, Effects of blue light on the circadian rhythm in *Acetabularia mediterranea*, *Protoplasma*, 105, 364, 1981.
15. **Schmid, R.**, Blue light effects on morphogenesis and metabolism, in *Acetabularia*, in *Blue Light Effects in Biological Systems*, Senger, H., Ed., Springer-Verlag, Berlin, 1984, 419.
16. **Schmid, R.**, Twofold effect of blue light on a circadian rhythm, in *Acetabularia*, *J. Interdisc. Cycle Res.*, in press.
17. **Schweiger, H.-G., Broda, H., and Wolff, D.**, Simultaneous recording of two circadian rhythms in an individual cell of *Acetabularia*, *Protoplasma*, 105, 365, 1981.
18. **Klemm, E. and Ninnemann, H.**, Detailed action spectrum for the delay shift in pupae emergence of *Drosophila pseudoobscura*, *Photochem. Photobiol.*, 24, 369, 1976.
19. **Winfree, A. T.**, Suppressing *Drosophila* circadian rhythm with dim light, *Science*, 183, 970, 1974.
20. **Lüning, K.**, Egg release in gametophytes of *Laminaria saccharina*. Induction by darkness and inhibition by blue light and UV, *Br. Phycol. J.*, 16, 379, 1981.
21. **Frank, K. D. and Zimmerman, W. F.**, Action spectra for phase shifts of a circadian rhythm in *Drosophila*, *Science*, 163, 688, 1969.
22. **Chandrashekaran, M. K. and Engelmann, W.**, Early and late subjective night phases of the *Drosophila pseudoobscura* circadian rhythm require different energies of blue light for phase shifting, *Z. Naturforsch.*, 28c, 750, 1973.
23. **Pittendrigh, C. S., Eichhorn, J. H., Minis, D. H., and Bruce, V. G.**, Circadian systems. IV. Photoperiodic time measurement in *Pectinophora gossypiella*, *Proc. Natl. Acad. Sci. U.S.A.*, 66, 758, 1970.
24. **Pittendrigh, C. S. and Minis, D. H.**, The photoperiodic time measurement in *Pectinophora gossypiella* and its relation to the circadian system in that species, in *Biochronometry*, Menaker, M., Ed., National Academy of Sciences, Washington, D.C., 1971, 212.
25. **Sargent, M. L. and Briggs, W. R.**, The effects of light on a circadian rhythm of conidiation in *Neurospora*, *Plant Physiol.*, 42, 1504, 1967.
26. **Schrott, E. L.**, The biphasic fluence response of carotenogenesis in *Neurospora crassa*: temporary insensitivity of the photoreceptor system, *Planta*, 151, 371, 1981.
27. **Feldman, J. F.**, Genetic approaches to circadian clocks, *Annu. Rev. Plant Physiol.*, 33, 583, 1982.
28. **Dharmananda, S.**, Studies on the circadian clock of *Neurospora crassa*: light induced phase shifting, Ph.D. thesis, University of California, Santa Cruz, 1980; as cited in **Feldman, J. F.**, *Plant Physiol.*, 33, 583, 1982.
29. **Deitzer, G. F., Horwitz, B. A., and Gressel, J.**, Circadian rhythm in the sensitivity of *Trichoderma harzianum* sporulation to induction by blue light, *Plant Physiol.*, 75 (Suppl.), 72, 1984.
30. **Halaban, R.**, Effects of light quality on the circadian rhythm of leaf movement of a short-day plant, *Plant Physiol.*, 44, 973, 1969.

31. **Aschoff, J.,** Exogenous and endogenous components in circadian rhythms, *Cold Spring Harbor Symp. Quant. Biol.,* 25, 11, 1960.
32. **Sargent, M. L., Briggs, W. R., and Woodward, D. O.,** Circadian nature of a rhythm expressed by an invertaseless strain of *Neurospora crassa, Plant Physiol.,* 41, 1343, 1966.
33. **Winfree, A. T.,** Integrated view of resetting a circadian clock, *J. Theor. Biol.,* 28, 327, 1970.
34. **Chandrashekaran, M. K. and Engelmann, W.,** Amplitude attenuation of the circadian rhythm in *Drosophila* with light pulses of varying irradiance and duration, *Int. J. Chronobiol.,* 3, 231, 1976.
35. **Winfree, A. T.,** Slow dark adaptation in *Drosophila's* circadian clock, *J. Comp. Physiol.,* 77, 418, 1972.
36. **Engelmann, W. and Honegger, H. W.,** Tagesperiodische Schlüpfrhythmik einer augenlosen *Drosophila melanogaster-* Mutante, *Naturwissenschaften,* 53, 588, 1966.
37. **Zimmerman, W. F. and Goldsmith, T. H.,** Photosensitivity of the circadian rhythm and the visual receptors in carotenoid-depleted *Drosophila, Science,* 1971, 1167, 1971.
38. **Klemm, E.,** Photorezeptor für Blaulicht bei *Drosophila pseudoobscura,* Master's thesis, University Tübingen, Tübingen, West Germany; as cited in Ninnemann, H., *Photochem. Photobiol. Rev.,* Vol. 4, Smith, K. C., Ed., Plenum Press, New York, 1979, 207.
39. **Zimmerman, W. F., Pittendrigh, C. S., and Pavlidis, T.,** Temperature compensation of the circadian oscillation in *Drosophilia pseudoobscura* and its entrainment by temperature cycles, *J. Insect Physiol.,* 14, 669, 1968.
40. **Hamm, U., Chandrashekaran, M. K., and Engelmann, W.,** Temperature sensitive events between photoreceptor and circadian clock?, *Z. Naturforsch.,* 30c, 240, 1975.
41. **Rau, W.,** Blue light-induced carotenoid biosynthesis in microorganisms, in *The Blue Light Syndrome,* Senger, H., Eds., Springer-Verlag, Berlin, 1980, 283.
42. **Klemm, E. and Ninnemann, H.,** Nitrate reductase — a key enzyme in blue light-promoted conidiation and absorbance change of *Neurospora, Photochem. Photobiol.,* 29, 629, 1979.
43. **Ninnemann, H. and Klemm-Wolfgramm, E.,** Blue light-controlled conidiation and absorbance change in *Neurospora* are mediated by nitrate reductase, in *The Blue Light Syndrome,* Senger, H., Ed., Springer-Verlag, Berlin, 1980, 238.
44. **Roldan, J. M. and Butler, W. L.,** Photoactivation of nitrate reductase from *Neurospora crassa, Photochem. Photobiol.,* 32, 375, 1980.
45. **Paietta, J. and Sargent, M. L.,** Blue light responses in nitrate reductase mutants of *Neurospora crassa, Photochem. Photobiol.,* 35, 853, 1982.
46. **Paietta, J. and Sargent, M. L.,** Photoreception in *Neurospora crassa:* correlation of reduced light sensitivity with flavin deficiency, *Proc. Natl. Acad. Sci. U.S.A.,* 78, 5573, 1981.
47. **Paietta, J. and Sargent, M. L.,** Modification of blue light photoresponses by riboflavin analogs in *Neurospora crassa, Plant Physiol.,* 72, 764, 1983.
48. **Brain, R. D., Freeberg, J., Weiss, C. V., and Briggs, W. R.,** Blue light-induced absorbance changes in membrane fractions from corn and *Neurospora, Plant Physiol.,* 59, 948, 1977.
49. **Muñoz, V. and Butler, W. L.,** Photoreceptor pigment for blue light responses in *Neurospora crassa, Plant Physiol.,* 55, 421, 1975.
50. **Britz, S. J., Schrott, E., Widell, S., Brain, R. D., and Briggs, W. R.,** Methylene blue-mediated red light photoreduction of cytochromes in particulate fractions of corn and *Neurospora, Carnegie Inst. Washington Yearb.,* 1976, 289, 1977.
51. **Brain, R. D., Woodward, D. O., and Briggs, W. R.,** Correlative studies of light sensitivity and cytochrome content in *Neurospora crassa, Carnegie Inst. Washington Yearb.,* 1976, 295, 1977.
52. **Klemm, E. and Ninnemann, H.,** Correlation between absorbance changes and a physiological response induced by blue light in *Neurospora, Photochem. Photobiol.,* 28, 227, 1978.
53. **Nakashima, H. and Feldman, J. F.,** Temperature sensitivity of light induced phase shifting of the circadian clock of *Neurospora, Photochem. Photobiol.,* 32, 247, 1980.
54. **Nakashima, H.,** Effects of membrane ATPase inhibitors on light-induced phase shifting of the circadian clock in *Neurospora crassa, Plant Physiol.,* 69, 619, 1982.
55. **Nakashima, H.,** Phase shifting of the circadian clock by diethylstilbestrol and related compounds in *Neurospora crassa, Plant Physiol.,* 70, 982, 1982.
56. **Nakashima, H. and Fujimura, Y.,** Light-induced phase shifting of the circadian clock in *Neurospora crassa* requires ammonium salts at high pH, *Planta,* 155, 431, 1982.
57. **Forward, A. B., Jr. and Davenport, D.,** The circadian rhythm of a behavorial photoresponse in the dinoflagellate *Gyrodinium dorsum, Planta,* 92, 259, 1970.
58. **Reinhardt, B.,** A rhythmic change in the enhancement of the dark resporation of *Chlorella fusca* induced by a short blue light exposure of low intensity, in *The Blue Light Syndrome,* Senger, H., Ed., Springer-Verlag, Berlin, 1980, 401.
59. **Sweeney, B. M. and Haxo, F. T.,** Persistence of photosynthetic rhythm in enucleated *Acetabularia, Science,* 134, 1361, 1961.

60. **Koop, H.-U., Schmid, R., Heunert, H.-H., and Milthaler, B.,** Chloroplast migration: a new circadian rhythm in *Acetabularia, Protoplasm,* 97, 301, 1978.
61. **Durand, R.,** Photoperiodic response of *Coprinus congregatus.* Effects of light breaks on fruiting, *Physiol. Plant.,* 55, 226, 1982.
62. **Dring, M. J. and Lüning, K.,** A photoperiodic response mediated by blue light in the brown alga *Scytosiphon lomentaria, Planta,* 125, 25, 1975.
63. **Hanke, J., Hartmann, K. M., and Mohr, H.,** Die Wirkung von "Störlicht" auf die Blütenbildung von *Sinapis alba L., Planta,* 86, 235, 1969.
64. **Ito, M.,** Light induced synchrony of cell division in the protonemata of the fern *Pteris vittata, Planta,* 90, 22, 1970.
65. **Ito, M.,** Effects of blue and red light on the cell division cycle in *Pteris* protonemata, *Plant Sci. Lett.,* 3, 351, 1974.
66. **Miyata, M., Wada, M., and Furuya, M.,** Effects of phytochrome and blue-near ultraviolet light-absorbing pigment on duration of component phases of the cell cycle in *Adiantum* gametophytes, *Dev. Growth Differ.,* 21, 577, 1979.
67. **Furuya, M., Wada, M., and Kadota, A.,** Regulation of cell growth and cell cycle by blue light in *Adiantum* gametophytes, in *The Blue Light Syndrome,* Senger, H., Ed., Springer-Verlag, Berlin, 1980, 119.
68. **Jerebzoff, S. and Jaques, R.,** Recherche du système pigmentaire actif sur la stabilisation du rythme interne de Leptosphaeria michotii, *C. R. Acad. Sci. Paris,* 268, 691, 1969.
69. **Njus, D., Sulzman, F. M., and Hastings, J. W.,** Membrane model for the circadian clock, *Nature (London),* 248, 116, 1974.
70. **Njus, D., Van Gooch, D., Mergenhagen, D., Sulzman, F., and Hastings, J. W.,** Membranes and molecules in circadian systems, *Fed. Proc.,* 35, 2353, 1976.
71. **Sweeney, B. M.,** A physiological model for circadian rhythms derived from the *Acetabularia* rhythm paradoxes, *Int. J. Chronobiol.,* 2, 25, 1974.
72. **Burgoyne, R. D.,** A model for the molecular basis of circadian rhythms involving monovalent ion-mediated translational control, *FEBS Lett.,* 94, 17, 1978.
73. **Schweiger, H.-G. and Schweiger, M.,** Circadian rhythms in unicellular organisms: an endeavour to explain the molecular mechanism, *Int. Rev. Cytol.,* 51, 315, 1977.

Chapter 9

MOVEMENT

Donat-P. Häder

TABLE OF CONTENTS

I. INTRODUCTION

This chapter will cover light-dependent movement from microorganisms and intracellular organelles such as chloroplasts. It will not deal with phototropism of higher or lower plants. Only a few motile microorganisms utilize a photoreceptor of the action spectrum which corresponds with that defined for cryptochrome.[1,2] Even some flagellates show an "extended blue-light" response. Therefore, not only flavin/carotenoid reactions are covered in this chapter. Some types of photoorientation, however, which are clearly controlled by the photosynthetic apparatus will be summarized in a separate chapter (see Chapter 11 by D.-P. Häder).

Microorganisms utilize external stimuli as clues to find their specific ecological niche in their microenvironment. Among the various stimuli such as chemical, thermal, magnetical, mechanical, and gravitational,[3] light plays a major role for most motile microorganisms. This is obvious for photosynthetic organisms which depend on the availability of visible — plus near ultraviolet (UV) and infrared (IR) — radiation. Nonphotosynthetic organisms also orient themselves with respect to light, for instance, to move to the surface where spores are produced or to move into shaded areas to avoid photooxidation of their intracellular pigments.

II. DEFINITION OF RESPONSES

The most obvious light-mediated response is *phototaxis*, which describes an orientation with respect to the light direction.[4] Some organisms show positive phototaxis (toward the light source, Figure 1a) or negative phototaxis (Figure 1b).[5-8] In some cases movement at an angle left and right of the light source has been observed (which is called dia- or transversal phototaxis in the case of a movement perpendicular to the light direction; (Figure 1c).[9,10]

Photokinesis is independent of the direction of light. It can be described as a steady-state dependence of the speed of movement on the light intensity.[4,11] A positive photokinesis is defined as a higher velocity at a given light intensity than in the dark control. Some ciliates do not move in darkness and are activated by light.[12-15] This behavior leads to an accumulation of organisms in shaded areas. Other organisms stop moving in high light intensities.

Sudden changes in light intensity cause *photophobic responses*. As is photokinesis, this reaction is independent of the light direction. Both a step-up or a step-down in light intensity can cause a response, and some organisms respond to either stimulus in the appropriate light-intensity range.[11] The intensity change has to occur suddenly rather than gradually and can be either temporal or spatial.[16] The type of motor reaction is a built-in function of each species: some flagellates turn sidewise or rotate on the spot,[16,17] while cyanobacteria reverse the direction of movement.[18]

Orientation of intracellular organelles with respect to light has on occasion also been called phototaxis. Though plastid movement can be controlled by the direction of light, the term "phototaxis" will be restricted to the directed movement of freely motile microorganisms.

III. PHOTOMOVEMENT OF MICROORGANISMS

A. Photoreceptor Pigments
1. Flagellates
While some microorganisms utilize the photosynthetic pigments as photoreceptor pigments, most flagellates use a specialized cryptochrome system for photoorientation. Many action spectra indicate the involvement of a flavin-type photoreceptor since they closely resemble the absorption spectrum of riboflavin (Figure 2).[19-21] Studies with polarized light in *Euglena* indicate that the photoreceptor molecules are oriented in a dichroic array.[22] The

Table 1
COMPARISON OF THE PHOTOBEHAVIOR OF SYMBIOTIC AND APOSYMBIOTIC CILIATES

Organism	Photophobic responses		Photoaccumulation	Photokinesis
	Step-up	Step-down		
C. virens				
Aposymbiotic	+	−	−	−
Symbiotic	+	−	−	−
E. daidaleos				
Aposymbiotic	+	−	−	−
Symbiotic	+	+	+	+
P. bursaria				
Aposymbiotic	+	−	−	−
Symbiotic	+	+	+	+
P. arcticum[a]				
Aposymbiotic	+	−	−	−

[a]   No symbiotic form available.

3. Slime Molds

The cellular slime mold *Dictyostelium discoideum* undergoes a life cycle including a multicellular stage, called a pseudoplasmodium or slug, and a unicellular stage where the organism forms individual amoebae.[63] The multicellular slugs show exclusively positive phototaxis[64-68] (or a bidirectional orientation right or left of the light source).[9,10,69] the action spectrum indicates the involvement of a high-spin heme protein.[64,65,70-72]

The amoebae show both positive and negative phototaxis, depending on the light intensity.[73-76] The action spectrum is significantly different from that of the slugs, which suggests a different photoreceptor.[73,74,77] Furthermore, a mutant, which does not show any photoorientation in the slug stage, orients perfectly well in the amoebal stage.[78] The photoreceptor has been proposed to be a protoporphyrin IX.[78a] Thus, we are faced with the interesting situation that the two stages in the developmental cycle of *Dictyostelium* utilize different (though biochemically related) photoreceptor molecules.

The acellular slime mold *Physarum polycephalum* moves by cytoplasmic streaming.[79] Light regulates the oscillatory shuttle movement which is based on the contraction of the plasmatic actin-myosin system.[80-84] The photoreceptor seems to be a typical blue light receptor which controls motility and phototactic orientation.[85-87]

4. Bacteria

The prokaryote *Halobacterium halobium* contains several pigments which are related to the rhodopsins serving as photoreceptor pigments in animal vision.[88-92] The bacterium responds to either a step-down in light intensity of green light (maximum at about 565 nm) or a step-up in light intensity of blue and UV radiation (maximum around 370 nm)[93-98] with a reversal of movement direction.[99] The original hypothesis assumed that the main pigment, bacteriorhodopsin, which acts as a light-driven proton pump also controls the step-down phobic responses. Indeed, *Halobacterium* responds to changes in $\Delta\psi_{H^+}$.[100] However, bacteriorhodopsin-deficient mutants have been isolated which are still capable of photosensory behavior.[101] These mutants contain halorhodopsin which is a light-dependent chloride pump.[102,103] Halorhodopsin-deficient mutants likewise respond to both step-up and -down stimuli, so that we are left with another rhodopsin as a potential photoreceptor which has been called "slow-cycling" rhodopsin. Upon irradiation it undergoes a cycle with several intermediates. The backreaction to SR_{587} can be either mediated by light (373 nm) or takes place in darkness with a rather long time constant of 0.9 sec (Figure 7). Thus, the same

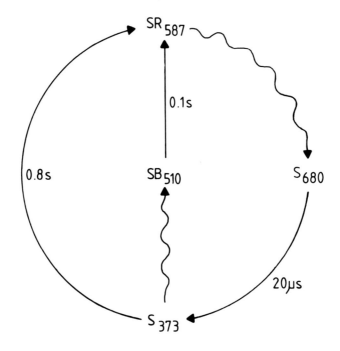

FIGURE 7.   Photochemical cycle of slow cycling rhodopsin upon light excitation.

molecule could be responsible for both light responses.[104] The alternative is that two separate photosystems exist which is supported by the observation that the two responses are developed at different times in the cell cycle.[105-107]

A number of other bacteria also show light-induced behavioral responses. Photosynthetic purple bacteria utilize the photosynthetic apparatus[108] (see Chapter 11, this volume). In *Salmonella* and *Escherichia* light-mediated responses seem to be rather unspecific because very high light intensities are required.[109] Possibly one of the flavins present in the cells doubles as a photoreceptor.

B. Photoreceptor Structures

The photoreceptor pigments of *Euglena* are concentrated in the paraflagellar body (PFB), a swelling at the base of the one emerging flagellum. It is located inside the reservoir and covered by the flagellar membrane (Figure 8). The second flagellum does not emerge from the reservoir, and the tip is attached to the longer flagellum near the PFB.[110,111]

In *Chlamydomonas* no PFB has been found. The photoreceptor seems to be localized in the vicinity of the stigma which is located on the equator of the almost spherical cell. The stigma consists of one or more layers of lipid droplets colored by carotenoids. It seems to play a role in light direction detection though it may not contain the actual photoreceptor. So-called eyeless mutants, which have a disordered stigma globule arrangement, are impaired in their phototactic orientation.[112] The photoreceptor molecules can be located in the membranes overlaying the stigma area. Both the outer chloroplast membrane and the plasmalemma have been considered as possible photoreceptor sites.[113,114] Freeze-fracture studies in a number of flagellates and green alga zoospores have revealed intramembranous particles; these are different in size and distribution in both membranes overlaying the stigma area than in any other portion of the cell (Figure 9).[115,116] These proteinaceous particles can be either photoreceptor particles or can play a role in the sensory transduction chain. Studies in many

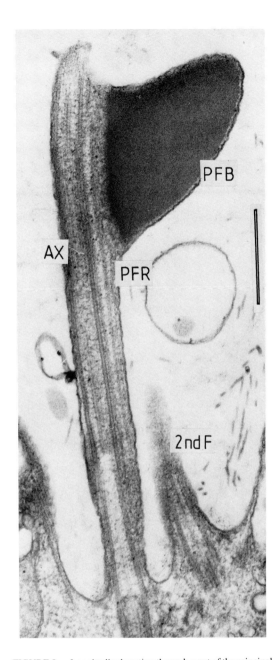

FIGURE 8. Longitudinal section through most of the principal flagellum of *E. mutabilis* inside the reservoir region. PFB, paraflagellar body; ax, axoneme; PFR, paraflagellar rod; 2ndF, second, minor flagellum. (Magnification × 48,000; bar represents 500 nm.) (From Häder, D.-P. and Melkonian, M., *Arch. Microbiol.*, 135, 25, 1983. With permission.)

other systems support the general statement that photoreceptors seem always to be part of or associated with membranes.[117-123]

C. Strategies for Direction Detection

Light-direction detection is not a trivial problem in single-celled or even multicellular microorganisms. Higher organisms utilize a complicated optical apparatus for this task. The

FIGURE 9.   Intramembranous particles in the (A) P face (magnification × 120,000) and (B) E face (magnification × 102,500) of the plasmalemma overlaying the stigma area of *Chlamydomonas*. (From Melkonian, M. and Robenek, H., *J. Ultrastruct. Res.*, 72, 90, 1980. With permission.)

FIGURE 9B

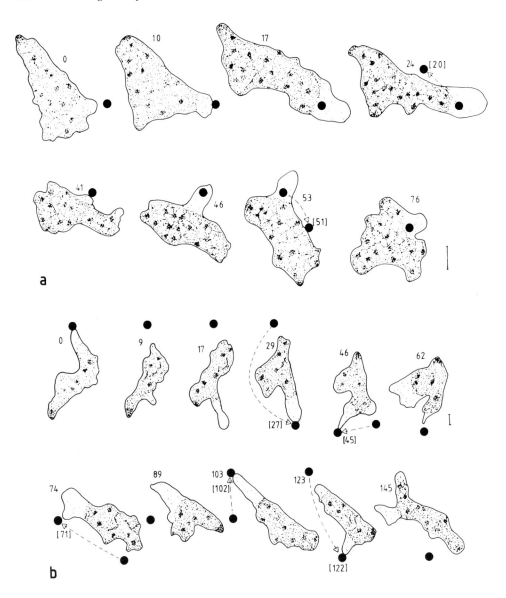

FIGURE 10.   Responses of *Dictyostelium* amoebae to stimulation by low and high illuminance microbeams of (A) 0.2 W · m^{-2} and (B) 1 × 10^4 W · m^{-2} (black spot). Numbers on top of the drawings indicate times in seconds at which pictures were taken. The numbers in the lines indicate times at which the microbeam was shifted into the new positions indicated by the arrows. Bars represent 2 μm. (From Häder, D.-P. et al., *Cell Biol. Int. Rep.*, 7, 611, 1983. With permission.)

mechanism of orientation can be analyzed only by direct observation of behavior. This rather tedious task can be facilitated by video analysis or[124-126] by semiautomatic or fully automatic computer tracking based on image analysis.[127-130] Motility can be analyzed by a laser velocimeter,[131] cinematographical techniques,[132] or with a fully automated computerized system.[133]

    There are two basic strategies for light-direction detection. The first mechanism is based on the simultaneous measurement of light intensity by more than one photoreceptor in an organism using a so-called *one-instant mechanism*.[134,135] By comparison of the simultaneous readings of several photoreceptors, the organism could detect an internal light gradient, caused by internal absorption or scattering, without leaving its spot. This spatial measurement seems to be utilized by slow, gliding organisms such as cyanobacteria, diatoms, or desmids. The coin-shaped desmid *Micrasterias* can turn its face to the light source while standing on its edge.[136,137]

An intracellular light gradient can be produced by two different phenomena. The light can be absorbed or scattered in the plasma or by cellular organelles so that the distal receptors measure less light intensity than the proximal ones.[138]

The existence of a one-instant mechanism can be proven by using a partial irradiation. A light spot is focused onto one portion of a cell while the remainder is in darkness.[139-142] Amoebae of the slime mold *Dictyostelium* can be induced to move in any given direction by placing a weak light spot on their perimeter, but they retreat from a bright light spot (Figure 10). A light intensity difference of only a few percent measured between the receptors may be sufficient for orientation.[143]

The alternative mechanism to produce an internal light gradient is to use a lens effect.[143] The internal refractive index of *Dictyostelium* (n = 1.37) is sufficiently higher than that of the surrounding medium (air, n = 1.0) to operate as a cylindrical lens.[144] Thus, during phototactic orientation, laterally impinging light is focused onto the distal side. The organism turns away from this brighter distal flank and by that mechanism moves towards the light source. The lens effect can be destroyed by incorporating an absorbing dye into the cells. In UV-B irradiation (< 300 nm), the slugs also turn away from the radiation source because of its naturally high internal absorption in this wavelength band.[145]

A second strategy is used by organisms which measure the light intensity at one point in time, store the value in a memory, and measure again a certain period of time later. By comparing two measurements, they could deduct whether they are moving up or down a light-intensity gradient. This temporal measurement strategy has been called a *two-instant mechanism* and seems to be used by fast-moving organisms like swimming flagellates or ciliates.[134,135] This strategy requires only one photoreceptor but it fails in parallel light in the absence of absorption in the medium since there is no intensity gradient.

This problem is overcome by a periodic shading mechanism. Many ciliates and flagellates move on a helical path and rotate simultaneously around their long axis.[146] When the flagellate *Euglena gracilis* swims in lateral light, the stigma casts a shadow on the photoreceptor (PFB) once during each rotation (Figure 11).[11,19,22,138] The ventral trailing flagellum swings away from the cell body in response to the shading which turns the front end towards the light source. Thus, repetitive photophobic responses are considered to be the basis for phototaxis in this organism. During negative phototaxis, the whole rear end of the cell is thought to act as an additional shading body.

This hypothetical mechanism may not be applied for all flagellates. *Chlamydomonas* also moves on a helical path which may be more or less pronounced. The geometrical properties of this flagellate prevent the stigma from casting a shadow over the flagellar basis. If the photoreceptor is located in the membrane area overlaying the stigma, not only this organelle, but also the whole cell body with its cup-shaped chloroplast shades the photoreceptor from the rear. Thus, not a periodic shading but a periodic irradiation — when the cell faces the light source — seems to be the signal.

A novel explanation has been proposed by Foster and Smyth.[40] Based on the observation that some stigmata consist of more than one layer of lipid droplets, they have suggested that this organelle could serve as an interference reflector. The individual aqueous and lipid layers have different refractive indices and are spaced at about 120 nm, which is about a quarter of the wavelength of maximal activity in, for example, *Chlamydomonas* (Figure 12).[47] An incoming light beam perpendicular to the surface is partially reflected at the transition between layers with different refractive indices. In fact, the light reflected from the stigma can be observed microscopically. The reflected wave constructively interferes with the incoming wave which causes a maximal signal near the plasmalemma. This hy-

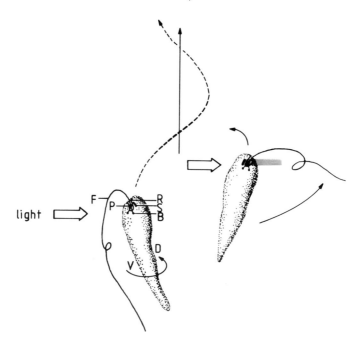

FIGURE 11.   Schematic drawing of *E. gracilis* orienting in a lateral light field. The stigma (S) casts a shadow over the paraflagellar body (P) when the dorsal side (D) faces the light. This causes the flagellum (F) originating in a basal body (B) inside the reservoir (R) to swing out to the ventral side (V). (From Tevini, M. and Häder, D.-P., in *Allgemeine Photobiologie,* Thieme, Stuttgart, 1985. With permission.)

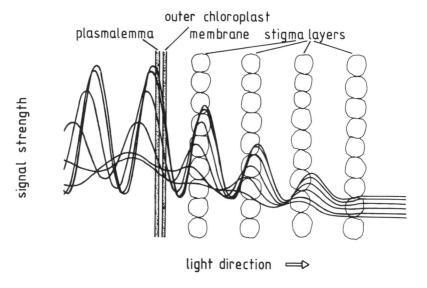

FIGURE 12.   Schematic representation of a quarter wavelength stack. Lightwaves hit the plasmalemma at angles of 90 (highest curve), 80, 70, 60, 50, and 40° (lowest curve) and are reflected at the interfaces between the lipid layers (n = 1.6) and aqueous layers (n = 1.35). The signal is amplified by constructive interference and has maximal strength close to the plasmalemma.[40]

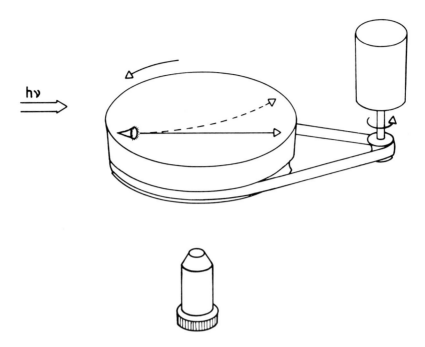

FIGURE 13. Path of a ciliate *(Stentor)* in a rotating cuvette orienting negative phototactically away from the light source (arrow). Up to a certain angular speed (10°/sec), the cells are able to correct their course.

pothesis is further supported by the observation of special intramembranous particles found in the plasmalemma overlying the stigma (see above).[114-116] Thus, the stigma/photoreceptor array could be envisioned as an antenna which scans the horizon during the helical rotation of the cell. The signal received by the photoreceptor is modulated when the cell rotates in lateral light. This modulation could serve as an error signal for course corrections, and will be zero when the cell faces the light source or moves away from it.

All the models discussed above are based on the idea that repetitive small course corrections cause a smooth turn towards (positive phototaxis) or away from (negative phototaxis) the light source. But this behavior was never observed. Therefore, Häder and Lebert[146a] used a different approach (Figure 13). The ciliate *Stentor* is observed in a circular cuvette while moving away from the light source. If the cell is capable of small course corrections it should swim in a straight path with respect to the light source even when the cuvette is rotated. Otherwise, the track should be curved. The experiment shows that in fact *Stentor* is capable of correcting its course when the angular speed does not exceed 10° per second.

D. Sensory Transduction

Once the primary light stimulus has been perceived, the signal is processed and amplified within the cell to finally control the motor apparatus.[147,148] These intracellular events can be analyzed only by indirect methods. For this purpose, the comparison of mutants,[149] the application of chemical inhibitors,[150-155] or the study of electrical phenomena[156,157] have proven to be successful approaches.

In several cases, the primary consequence of photoperception seems to be a vectorial proton transport. When the photoreceptor molecule is a membrane-bound protein, it can pick up a proton at one side of the membrane upon excitation and release it subsequently at the other. This phenomenon is well known in *Halobacterium*[89,96] and has been discussed as the primary event in photoperception of *Stentor* where a proton extrusion has been found during irradiation.[60,61,158,159] Proton gradients can be converted into ionic gradients which

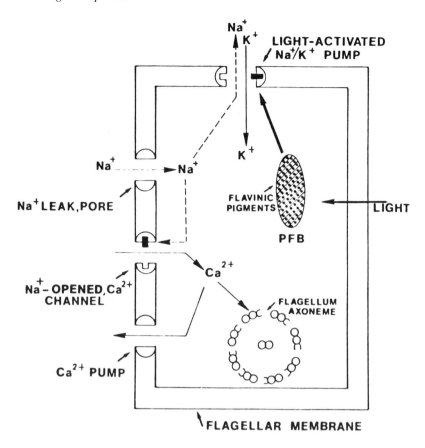

FIGURE 14.   Model for sensory transduction in the flagellate *Euglena gracilis*.

then act as secondary buffers.[160,161] Transport of ions across membranes not only changes the concentration of a given cellular compartment but also alters the electrical properties such as the resting potential of the cytoplasmic compartment with respect to the exterior.[162] Thus, electrical signals can be involved in the sensory transduction chain.[156,157] In *Haematococcus*, light-induced changes in the membrane potential can be measured only in the membrane segment overlaying the stigma.[163] In *Chlamydomonas*, the direction of the flagellar beat can be influenced by passing electrical currents into the cell by inserted microelectrodes.[164] Another example of electrical potential changes involved in the sensory transduction chain will be discussed in more detail in Chapter 11 of this volume.

Among the ions tested, calcium seems to play a major role in the sensory transduction chains of many organisms.[165-171] When *Cryptomonas* is cultured in the absence of calcium, it loses its ability to orient phototactically.[172] Calcium has also been shown to control the contraction of the flagellar roots in flagellates.[173] In *Chlamydomonas*, light modulates the activity of the flagella. A change in the light intensity causes a transient change in the beat pattern.[174] The two flagella of this organism respond differently to submicromolar levels of calcium,[175] which could thus cause a controlled movement with respect to light.

Diehn and Doughty have discussed a model for the sensory transduction in *Euglena* (Figure 14).[176-179] The primary perception of a photon in the PFB is thought to activate a Na^+/K^+ pump in the flagellar membrane. The intraflagellar sodium concentration controls Ca^{2+} channels in the flagellar membrane, which allow an influx of Ca^{2+} when opened. The increased Ca^{2+} concentration in turn effects a change in the flagellar beating pattern.

In some systems, cyclic nucleotides seem to act as secondary messengers in the sensory transduction chains. Their role was first analyzed in the chemotaxis of slime molds but has recently also been investigated in other systems. Some prokaryotes have been shown to utilize the methylation system found in bacterial chemotaxis also in light responses.[180-185]

The final step in the sensory transduction of motor responses of flagellates and ciliates is the reorientation of flagella and cilia. Electron microscopic studies have shown structural links between the stigma area and the flagellar basis by a system of root fibrils.[186] The three-dimensional analysis of the beat pattern[187,188] indicates that in addition to a modulation of the beat frequency a change in the beat direction can be initiated. This is caused by a rotation of the central pair of microtubuli in the flagellum.[189-191]

E. Ecological Significance of Photomovement

The necessity for an orientation in a light field is obvious for photosynthetic organisms. All three photoresponses can be used for this purpose. Obviously a number of different photoreceptor systems has been developed by multiple parallel evolution.[192-197] A positive phototaxis will bring an organism closer to the light source. This mechanism is used by flagellates to move upward in a column of water and get closer to the surface. This behavior leads to a vertical movement of planktonic organisms during the day/night cycle. It has been shown that even moonlight intensity influences the vertical migration pattern of demersal plankton.[198]

Also nonphotosynthetic organisms use phototaxis to move to the surface. Pseudoplasmodia of the cellular slime mold *Dictyostelium* are exclusively postively phototactic. This is of an ecological advantage for the organism since the spores will be developed at the surface where they can be distributed easily by wind and animals. Since the amoebae (which eventually develop into pseudoplasmodia) feed on bacteria in the forest mulch, not much light will be available to guide the organisms to the surface.[63] In fact, extremely low light intensities of 10^{-3} lx cause a perfect positive phototactic orientation.[70]

Some organisms have been shown to be affected by strong white light which could photobleach the cellular pigments.[14] Cyanobacteria, for example, are typical low-light-intensity organisms and thrive best at a few hundred lux. Higher intensities bleach the cells (Figure 15) and eventually kill the population within a few days.[199] In order to avoid the exposure to strong white light, some organisms (especially flagellates) use negative phototaxis. The delicate balance between positive phototaxis to low light intensities and negative phototaxis to high intensities causes the organisms to populate specific bands in the body of water which will move up and down with the changes in the light intensity during the day.

While cyanobacteria may use a phototactic orientation to escape from under the sediment which covers the population (occasionally caused by turbulences in the water),[200,201] they rely on photophobic responses to populate their suitable light-intensity niche.[202-205] When *Phormidium* moves into a shaded area it reverses its direction of movement and stays in the light by this simple but effective mechanism. The organism can even discriminate between light fields of only 4% difference in intensity.[206] Recently a step-up photophobic response was discovered by which the organism escapes from light intensities that are too bright.[199] The extremely sensitive discrimination by which the organisms select a suitable light intensity can be demonstrated by projecting a photographic negative into a preparation of cyanobacteria. The filaments accumulate in areas of appropriate intensities, discriminating between even subtle differences, and form a photographic positive (Figure 16). Colored ciliates like *Blepharisma* seem to rely on positive photokinesis to avoid bright light fields.[14] Symbiontic systems, on the other hand, accumulate in light fields, which is a necessity for the algal partner, using phobic responses.[15]

The orientation of microorganisms in light is easily disturbed by low doses of UV-B radiation (280 to 320 nm). UV-B has been shown to inhibit the motility of slime molds,

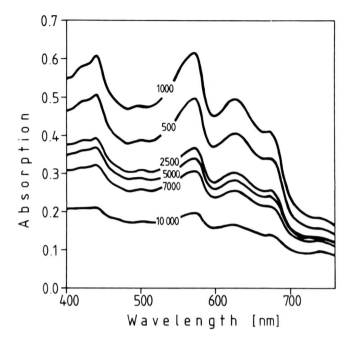

FIGURE 15.   Absorption spectrum of *Phormidium uncinatum* measured after 24-hr white light (illuminance in lux) irradiation.

prokaryotic and eukaryotic, and gliding and swimming algae.[199,207,208] It also affects the development of microorganisms.[209] The delicate balance between positive and negative phototaxis or between step-up and step-down phobic responses can be disturbed when one response is more inhibited than the other. This may result in a potentially harmful effect on the survival of populations of microorganisms if the current predictions concerning the decrease of the stratospheric ozone layer and resulting increase of UV-B radiation due to the emission of manmade chlorinated fluorocarbons materializes.[210] The microorganisms tested so far are not immediately harmed by the predicted UV-B levels but would die either from lack of light energy (when the response by which they are prevented from moving into shaded areas is impaired by UV) or by photooxidation of their pigments (when they are not prevented from moving into light fields that are too bright.)

IV. PHOTOORIENTATION OF PLASTIDS

Intracellular movement of cytoplasm and organelles has been observed in many microorganisms, plants, and animals.[211-215] In many cases this is under the control of light. Two types can be distinguished, though the separation may be an artificial one since intermediate types exist. The first is an effect of light on the cytoplasmic streaming with an alteration of the chloroplast arrangement as a consequence of the light-induced plasma movement. The second type is an oriented movement of plastids with respect to the light direction.[216,217]

A. Chloroplast Movement as a Result of Cytoplasmic Streaming

Mesophyll cells of *Vallisneria* show a steady rotation of the cytoplasma including the chloroplasts. After prolonged dark periods, the outer layers, which carry the chloroplasts, stop. Light induces movement of these outer layers, which is called *photodinesis*. The action spectrum indicates the activity of a typical blue light receptor. In addition, there is a secondary peak in the red which is attributed to chlorophyll. Centrifugation experiments suggest that the chloroplasts are attached to the cortical plasma layer (by fibrils?) and are loosened under the effect of light. This allows the plastids to be carried by the moving endoplasmic layers.[218]

FIGURE 16. Cyanobacteria *(Phormidium)* use photophobic responses to accumulate in areas of suitable light intensities and thus generate a positive of a photographic negative (castle in Marburg).

Cytoplasmic rotation is also found in the Characeae.[219,220] *Nitella* is a suitable object since the large internodal cells are not covered by cortical cells.[221-223] Only the endoplasmic layer is moving while the ectoplasm, which includes the plastids, is stationary. The driving force for this movement is produced by actin fibrils which have been demonstrated by electron microscopy as well as decorated heavy meromyosin. The actin filaments are stationary and convey the myosin molecules past them.[224-226]

In green coenocytic algae *(Bryopsis, Vaucheria)*, chloroplast movement is the consequence of light-dependent cytoplasmic streaming.[227,228] In darkness, the plastids are randomly distributed. When the thallus is kept in air, the plastids move to the wall segments more or less perpendicular to the light beam in low light intensities. In higher light intensities they populate the side walls (Figure 17). Again actin filaments are supposed to be the driving mechanism for cytoplasmic streaming, since cytochalasin B stops the movement. Since the chloroplasts are located in the endoplasm they participate in the movement. Experiments with microbeams have demonstrated that motility is halted by blue light.[229] The chloroplasts are carried along actin cables which disintegrate under the influence of light and form a network. This stops the movement of the chloroplasts which accumulate in the light spot.

A blue microbeam causes an outward electrical current at the irradiated spot in *Vaucheria* which causes a hyperpolarization of the membrane. This current is carried by a proton efflux, thus it has been suggested to be based on a light-dependent proton pump.[230] The action spectrum indicates the involvement of a typical cryptochrome which seems to be membrane bound because of a prominent action dichroism also found in other systems.[231-233]

dark   weak light   strong light

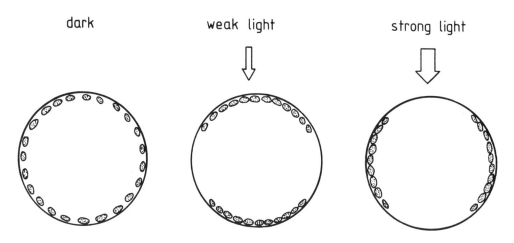

FIGURE 17.   Chloroplast distribution of *Vaucheria* in darkness and weak and strong light.[233]

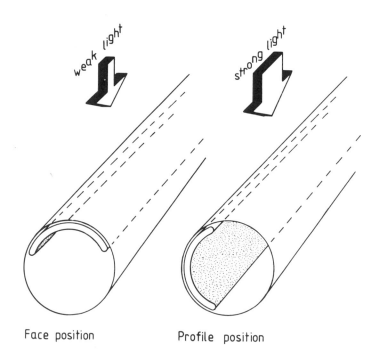

Face position   Profile position

FIGURE 18.   Chloroplast orientation of *Hormidium* in weak and strong light.[233]

B. Active Orientation of Chloroplasts

In many eukaryotic algae as well as aquatic and terrestrial plants, the chloroplasts line the inner cell walls perpendicular to the light beam (periclinal) in low light intensity and those parallel to the light beam (anticlinal) in higher intensities.[234-237] This arrangement can be disturbed by a circadian rhythm which is also found in *Acetabularia, Ulva,* and *Pyrocystis.*[238-243] There is a remarkable action dichroism in *Dictyota* similar to that in mosses and some higher plants. In low light intensity, the plastids move to the perclinal and those anticlinal walls parallel to the E-vector of polarized light. This is in agreement with the assumed orientation of the photoreceptor molecules parallel to the surface. No dichroism has been found for the high-intensity reaction.

The green algae *Hormidium* and *Mougeotia* contain only one large chloroplast per cell, which moves into a face position in low light intensity and into a profile position in high light intensity (Figure 18).[244-247] In air a *Hormidium* trichome acts as a cylindrical lens and focuses the light onto the distal wall where most chloroplasts move in low light intensity. In water — where the difference between the refractive indices inside and outside the cell is insignificant — the chloroplasts lie near either the proximal or distal wall. Judging from FMN starvation experiments and quencher studies (KI), flavins are favored as photoreceptors.[215,233,248,249] In addition, chlorophyll has been discussed as a second photoreceptor.

In *Mougeotia* and *Mesotaenium,* phytochrome has been shown to be the predominant photoreceptor[250-255] in addition to a blue light receptor.[256] Since the receptors are dichroic and parallel to the membrane, the edges of the chloroplast move to cell areas of lowest light absorption during the weak light orientation. The response can be induced even by microsecond light pulses. The profile position in strong light, however, cannot be induced by a short pulse, but rather continuous light or repetitive pulses have to be applied. The photoreceptor for this response seems to be typical blue light receptor.

The sensory transduction chain involves Ca^{2+}, which has been demonstrated by chlorotetracycline fluorescence to be localized in vesicles along the edge of the chloroplast.[257] The perception of light is supposed to be a trigger signal to liberate Ca^{2+} from the vesicles. The calcium binds to calmodulin and activates the myosin light chain kinase. The activated myosin in turn moves along the actin cables.[258,259] The orientation can be inhibited by calcium depletion and the movement by cytochalasin B.

REFERENCES

1. **Senger, H.,** Cryptochrome, some terminological thoughts, in *Blue Light Effects in Biological Systems,* Senger, H., Ed., Springer-Verlag, Berlin, 1984, 72.
2. **Briggs, W. R. and Iino, M.,** Blue-light-absorbing photoreceptors in plants, *Phil. Trans. R. Soc. London, Ser. B,* 303, 347, 1983.
3. **Spormann, A. M. and Wolfe, R. S.,** Chemotactic, magnetotactic and tactile behaviour in a magnetic *Spirillum, FEMS Microbiol. Lett.,* 22, 171, 1984.
4. **Diehn, B., Feinleib, M., Haupt, W., Hildebrand, E., Lenci, F. and Nultsch, W.,** Terminology of behavioral responses of motile microorganisms. *Photochem. Photobiol.,* 26, 559, 1977.
5. **Häder, D.-P., Colombetti, G., Lenci, F., and Quaglia, M.,** Phototaxis in the flagellates, *Euglena gracilis* and *Ochromonas danica, Arch. Microbiol.,* 130, 78, 1981.
6. **Haupt, W.,** Photoperception and photomovement, *Phil. Trans. Roy. Soc. London, Ser. B,* 303, 467, 1983.
7. **Feinleib, M. E.,** Photomotile responses in flagellates, in *Photoreception and Sensory Transduction in Aneural Organisms,* Lenci, F. and Colombetti, G., Eds., Plenum Press, New York, 1980, 45.
8. **Poff, K. L. and Hong, C. B.,** Photomovement and photosensory transduction in microorganisms, *Photochem. Photobiol.,* 36, 749, 1982.
9. **Fisher, P. R. and Williams, K. L.,** Bidirectional phototaxis by *Dictyostelium discoideum* slugs, *FEMS Microbiol. Lett.,* 12, 87, 1981.
10. **Fisher, P. R. and Williams, K. L.,** Activated charcoal and orientation behaviour by *Dictyostelium discoideum* slugs, *J. Gen. Microbiol.,* 126, 519, 1981.
11. **Häder, D.-P.,** Photomovement, in *Encyclopedia of Plant Physiology,* New Series, Vol. 7, *Physiology of Movements,* Haupt, W. and Feinleib, M. E., Eds., Springer-Verlag, Berlin, 1979, 268.
12. **Pado, R.,** The effect of white light on kinesis in the protozoans *Paramecium bursaria, Acta Protozool.,* 14, 83, 1975.
13. **Kraml, M. and Marwan, W.,** Photomovement responses of the heterotrichous ciliate *Blepharisma japonicum, Photochem. Photobiol.,* 37, 313, 1983.
14. **Matsuoka, T.,** Distribution of photoreceptors inducing ciliary reversal and swimming acceleration in *Blepharisma japonicum, J. Exp. Zool.,* 225, 337, 1983.
15. **Reisser, W. and Häder, D.-P.,** Role of endosymbiontic algae in photokinesis and photophobic responses of ciliates, *Photochem. Photobiol.,* 39, 673, 1984.
16. **Shimmen, T.,** Quantitative studies on step-down photophobic response of *Euglena* in an individual cell, *Protoplasma,* 106, 37, 1981.

17. **Diehn, B.,** Experimental determination and measurement of photoresponses, in *Photoreception and Sensory Transduction in Aneural Organisms,* Lenci, F. and Colombetti, G., Eds., Plenum Press, New York, 1980, 107.

18. **Häder, D.-P.,** Photosensory transduction chains in procaryotes, in *Photoreception and Sensory Transduction in Aneural Organisms,* Lenci, F. and Colombetti, G., Eds., Plenum Press, New York, 1980, 355.

19. **Diehn, B. and Kint, B.,** The flavin nature of the photoreceptor molecule for phototaxis in *Euglena, Physiol. Chem. Phys.,* 2, 483, 1970.

20. **Schmidt, W.,** Physiological bluelight reception, *Struct. Bonding (Berlin), 41,* 1, 1980.

21. **Hemmerich, P. and Schmidt, W.,** Bluelight reception and flavin photochemistry, in *Photoreception and Sensory Transduction in Aneural Organisms,* Lenci, F. and Colombetti, G., Eds., Plenum Press, New York, 1980, 271.

22. **Diehn, B.,** Two perpendicularly oriented pigment systems involved in phototaxis of *Euglena, Nature (London),* 122, 366, 1969.

23. **Suzaki, T. and Williamson, R. E.,** Photoresponse of a colorless euglenoid flagellate, *Astasia longa, Plant Sci. Lett.,* 32, 101, 1983.

24. **Mikolajczyk, E.,** Photophobic responses in *Euglenina* II. Sensitivity to light of the colorless flagellate *Astasia longa* in low and high viscosity medium, *Acta Protozool.,* 23, 85, 1984.

25. **Diehn, B.,** The receptor/effector system of phototaxis in *Euglena, Acta Protozool.,* 11, 325, 1972.

26. **Heelis, D. V., Heelis, P. F., Bradshaw, F., and Phillips, G. O.,** Does the stigma of *Euglena gracilis* play an active role in the photoreception processes of this organism? A photochemical investigation of isolated stigmata, *Photobiochem. Photobiophys.,* 3, 77, 1981.

27. **Piccinni, E. and Mammi, M.,** Motor apparatus of *Euglena gracilis:* ultrastructure of the basal portion of the flagellum and the paraflagellar body, *Boll. Zool.,* 45, 405, 1978.

28. **Robenek, H. and Melkonian, M.,** Structure specialization of the paraflagellar membrane of *Euglena, Protoplasma,* 117, 154, 1983.

29. **Colombetti, G. and Lenci, F.,** Identification and spectroscopic characterization of photoreceptor pigments, in *Photoreception and Sensory Transduction in Aneural Organisms,* Lenci, F. and Colombetti, G., Eds., Plenum Press, New York, 1980, 173.

30. **Banchetti, R., Rosati, G., and Verni, F.,** Cytochemical analysis of the photoreceptor in *Euglena gracilis* Klebs (Flagellata, Euglenoidina), *Monit. Zool. Ital.,* 14, 165, 1980.

31. **Benedetti, P. A. and Lenci, F.,** *In vivo* microspectrofluorometry of photoreceptor pigments in *Euglena gracilis, Photochem. Photobiol.,* 26, 315, 1977.

32. **Doughty, M. J. and Diehn, B.,** Flavins as photoreceptor pigments for behavioral responses in motile microorganisms, especially in the flagellated alga, *Euglena* sp., *Struct. Bonding (Berlin),* 41, 45, 1980.

33. **Vierstra, R. D. and Poff, K. L.,** Mechanism of specific inhibition of phototropism by phenylacetic acid in corn seedling, *Physiol. Plant.,* 67, 1011, 1981.

34. **Colombetti, G., Häder, D.-P., Lenci, F., and Quaglia, M.,** Phototaxis in *Euglena gracilis:* effect of sodium azide and triphenylmethylphosphonium ion on the photosensory transduction chain, *Curr. Microbiol.,* 7, 281, 1982.

35. **Lenci, F., Colombetti, G., and Häder, D.-P.,** Role of flavin quenchers and inhibitors in the sensory transduction of the negative phototaxis in the flagellate, *Euglena gracilis, Curr. Microbiol.,* 9, 285, 1983.

36. **Otto, M. K., Jayaram, M., Hamilton, R. M., and Delbrück, M.,** Replacement of riboflavin by an analogue in the blue-light photoreceptor of *Phycomyces, Proc. Natl. Acad. Sci.,* U.S.A., 78, 266, 1981.

37. **Ghisla, S., Mack, R., Blankenhorn, G., Hemmerich, P., Krienitz, E., and Kuster, T.,** Structure of a novel flavin chromophore from *Avena* coleoptiles, the possible "blue light" photoreceptor, *Eur. J. Biochem.,* 138, 339, 1984.

38. **Häder, D.-P. and Melkonian, M.,** Phototaxis in the gliding flagellate, *Euglena mutabilis, Arch. Microbiol.,* 135, 25, 1983.

39. **Halldal, P.,** Light and microbial activities, in *Contemporary Microbial Ecology,* Ellwood, D. C., Hedger, J. N., Latham, M. J., Slater, J. H., and Lynch, J. M., Eds., Academic Press, London, 1980, 1.

40. **Foster, K. W. and Smyth, R. D.,** Light antennas in phototactic algae, *Microbiol. Rev.,* 44, 572, 1980.

40a. **Foster, K. W., Saranak, J., Patel, N., Zarilli, G., Okabe, M., Kline, T., and Nakanishi, K.,** A rhodopsin is the functional photoreceptor for phototaxis in the unicellular eukaryote *Chlamydomonas, Nature (London),* 311, 756, 1984.

41. **Häder, D.-P.,** Photomovement, in *Blue Light Effects in Biological Systems,* Senger, H., Ed., Springer-Verlag, Berlin, 1984, 435.

42. **del Portillo, H. A. and Dimock, R. V., Jr.,** Specificity of the host-induced negative phototaxis of the symbiotic water mite, *Unionicola formosa, Biol. Bull.,* 162, 163, 1982.

43. **Buchanan, C. and Goldberg, B.,** The action spectrum of *Daphnia magna* (Crustacea) phototaxis in a simulated natural environment, *Photochem. Photobiol.,* 34, 711, 1981.

44. **Hertel, H.,** Phototaktische Reaktion von *Asplanchna priodonta* bei monochromatischem Reizlicht, *Z. Naturforsch.,* 34c, 148, 1979.

45. **Bensasson, R. V.,** Molecular aspects of photoreceptor function: carotenoids and rhodopsins, in *Photoreception and Sensory Transduction in Aneural Organisms,* Lenci, F. and Colombetti, G., Eds., Plenum Press, New York, 1980, 211.

46. **Forward, R., Jr.,** Phototaxis by the dinoflagellate *Gymnodinium splendens* Lebour, *J. Protozool.,* 21, 312, 1974.

47. **Nultsch, W., Throm, G., and Rimscha, J.,** Phototaktische Untersuchungen an *Chlamydomonas reinhardii* Dangeard in homokontinuierlicher Kultur, *Arch. Microbiol.,* 80, 351, 1971.

48. **Watanabe, M. and Furuya, M.,** Action spectrum of phototaxis in a cryptomonad alga, *Cryptomonas sp., Plant Cell Physiol.,* 15, 413, 1974.

49. **Iwatsuki, K. and Naitoh, Y.,** Photoresponses in colorless *Paramecium, Experientia,* 38, 1453, 1982.

50. **Iwatsuki, K. and Naitoh, Y.,** Behavioral responses in *Paramecium multimicronucleatum* to visible light, *Photochem. Photobiol.,* 37, 415, 1983.

51. **Reisser, W.,** Host-symbiont interaction in *Paramecium bursaria:* physiological and morphological features and their evolutionary significance, *Ber. Dtsch. Bot. Ges.,* 94, 557, 1982.

52. **Pado, R.,** Spectral activity of light and phototaxis in *Paramecium bursaria, Acta Protozool.,* 11, 387, 1972.

53. **Niess, D., Reisser, W., and Wiessner, W.,** The role of endosymbiotic algae in photoaccumulation of green *Paramecium bursaria, Planta,* 152, 268, 1981.

54. **Niess, D., Reisser, W., and Wiessner, W.,** Photobehaviour of *Paramecium bursaria* with different symbiotic and aposymbiotic species of *Chlorella, Planta,* 156, 475, 1982.

55. **Matsuoka, T.,** Negative phototaxis in *Blepharisma japonicum, J. Protozool.,* 30, 409, 1983.

56. **Song, P.-S., Häder, D.-P., and Poff, K. L.,** Step-up photophobic response in the ciliate, *Stentor coeruleus, Arch. Microbiol.,* 126, 181, 1980.

57. **Song, P.-S., Häder, D.-P., and Poff, K. L.,** Phototactic orientation by the ciliate, *Stentor coeruleus, Photochem. Photobiol.,* 32, 781, 1980.

58. **Kim, I.-H., Prusti, R. K., Song, P.-S., Häder, D.-P., and Häder, M.,** Phototaxis and photophobic responses in *Stentor coeruleus.* Action spectrum and role of Ca^{2+} fluxes, *Biochim. Biophys. Acta,* 779, 298, 1984.

59. **Song, P.-S., Walker, E. B., and Yoon, M. J.,** Molecular aspects of photoreceptor function in *Stentor coeruleus,* in *Photoreception and Sensory Transduction in Aneural Organisms,* Lenci, F. and Colombetti, G., Eds., Plenum Press, New York, 1980, 241.

60. **Song, P.-S., Walker, E. B., Auerbach, R. A., and Robinson, G. W.,** Proton release from *Stentor* photoreceptors in the excited states, *Biophys. J.,* 35, 551, 1981.

61. **Song, P.-S.,** Photosensory transduction in *Stentor coeruleus* and related organisms, *Biochim. Biophys. Acta,* 639, 1, 1981.

62. **Tartar, V.,** Caffeine bleaching of *Stentor coeruleus, J. Exp. Zool.,* 181, 245, 1972.

63. **Poff, K. L. and Whitaker, B. D.,** Movement of slime molds, in *Encyclopedia of Plant Physiology,* N.S. Vol. 7., *Physiology of Movements,* Haupt, W. and Feinleib, M. E., Eds., Springer-Verlag, Berlin, 1979, 355.

64. **Bonner, J. T., Clarke, W. W., Jr., Neely, C. L., Jr., and Slifkin, M. K.,** The orientation to light and the extremely sensitive orientation to temperature gradients in the slime mold *Dictyostelium discoideum, J. Cell. Comp. Physiol.,* 36, 149, 1950.

65. **Francis, D. W.,** Some studies on phototaxis in *Dictyostelium, J. Cell. Comp. Physiol.,* 64, 131, 1964.

66. **Fisher, P. R., Smith, E., and Williams, K. L.,** An extracellular chemical signal controlling phototactic behavior by *Dictyostelium discoideum* slugs, *Cell,* 23, 799, 1981.

67. **Häder, D.-P. and Burkart, U.,** Mathematical simulation of *Dictyostelium* pseudoplasmodia movements, *Math. Biosci.,* 67, 41, 1983.

68. **Smith, E., Fisher, P. R., Grant, W. N., and Williams, K. L.,** Sensory behaviour in *Dictyostelium discoideum* slugs: phototaxis and thermotaxis are not mediated by a change in slug speed, *J. Cell Sci.,* 54, 329, 1982.

69. **Häder, D.-P. and Burkart, U.,** Movement of *Dictyostelium* pseudoplasmodia *in vivo* and in a mathematical simulation, *Plant Cell Physiol.,* 25, 705, 1984.

70. **Poff, K. L. and Häder, D.-P.,** An action spectrum for phototaxis by pseudoplasmodia of *Dictyostelium discoideum, Photochem. Photobiol.,* 39, 433, 1984.

71. **Poff, K. L. and Butler, W. L.,** Spectral characteristics of the photoreceptor pigment of phototaxis in *Dictyostelium discoideum, Photochem. Photobiol.,* 20, 241, 1974.

72. **Poff, K. L., Loomis, W. F., Jr., and Butler, W. L.,** Isolation and purification of the photoreceptor pigment associated with phototaxis in *Dictyostelium discoideum, J. Histochem. Cytol.,* 249, 2164, 1974.

73. **Häder, D.-P. and Poff, K. L.,** Light-induced accumulations of *Dictyostelium discoideum* amoebae, *Photochem. Photobiol.,* 29, 1157, 1979.

74. **Häder, D.-P. and Poff, K. L.,** Photodispersal from light traps by amoebas of *Dictyostelium discoideum, Exp. Mycol.,* 3, 121, 1979.

75. **Hong, C. B.,** Thermosensing and Photosensing in *Dictyostelium discoideum* Amoebae, Ph.D. thesis, Michigan State University, East Lansing, 1983.
76. **Häder, D.-P., Williams, K. L., and Fisher, P. R.,** Phototactic orientation by amoebae of *Dictyostelium discoideum* slug phototaxis mutants, *J. Gen. Microbiol.,* 129, 1617, 1983.
77. **Hong, C. B., Häder, M. A., Häder, D.-P., and Poff, K. L.,** Phototaxis in *Dictyostelium discoideum* amoebae, *Photochem. Photobiol.,* 33, 373, 1981.
78. **Häder, D.-P., Whitaker, B. D., and Poff, K. L.,** Responses to light by a nonphototactic mutant of *Dictyostelium discoideum, Exp. Mycol.,* 4, 382, 1980.
78a. **Poff, K. L.,** personal communication, 1984.
79. **Britz, S. J.,** Cytoplasmic streaming in *Physarum,* in *Encyclopedia of Plant Physiology,* N.S., Vol. 7, *Physiology of Movements,* Haupt, W. and Feinleib, M. E., Eds., Springer-Verlag, Berlin, 1979, 127.
80. **Wohlfarth-Bottermann, K. E. and Block, I.,** Function of cytoplasmic flow in photosensory transduction and phase regulation of contractile activities in *Physarum, Cold Spring Harbor Symp. Quant. Biol.,* 46, 563, 1982.
81. **Wohlfarth-Bottermann, K. E. and Block, I.,** The pathway of photosensory transduction in *Physarum polycephalum, Cell Biol. Int. Rep.,* 5, 365, 1981.
82. **Hatano, S., Sugino, H., and Ozaki, K.,** Regulatory proteins of actin polymerization from *Physarum* plasmodium and classification of actin-associated proteins, in *Actin: Structure and Function in Muscle and Non-muscle Cells,* Remedios, D., Ed., Academic Press, Australia, 1983, 277.
83. **Baranowski, Z., Shraideh, Z., and Wohlfarth-Bottermann, K. E.,** Which phase of the contraction-relaxation cycle of cytoplasmic actomyosin in *Physarum* is modulated by blue light?, *Cell Biol. Int. Rep.,* 6, 859, 1982.
84. **Wohlfarth-Botterman, K. E. and Block, I.,** Function of cytoplasmic flow in photosensory transduction and phase regulation of contractile activities in *Physarum, Cold Spring Harbor Symp. Quant. Biol.,* 46, 563, 1981.
85. **Block, I. and Wohlfarth-Bottermann, K. E.,** Blue light as a medium to influence oscillatory contraction frequency in *Physarum, Cell Biol. Int. Rep.,* 5, 73, 1981.
86. **Schreckenbach, T. and Verfuerth, C.,** Blue light influences gene expression and motility in starving microplasmodia of *Physarum polycephalum, Eur. J. Cell Biol.,* 28, 12, 1982.
87. **Häder, D.-P. and Schreckenbach, T.,** Phototactic orientation in plasmodia of the acellular slime mold, *Physarum polycephalum, Plant Cell Physiol.,* 25, 55, 1984.
88. **Hildebrand, E.,** Halobacteria: the role of retinalprotein complexes, in *The Biology of Photoreception,* Cosens, D. J. and Vince-Price, D., Eds., Society for Experimental Biology Symposium No. 36, Cambridge University Press, Cambridge, 1983, 207.
89. **Stoeckenius, W. and Bogomolni, R. A.,** Bacteriorhodopsin and related pigments of halobacteria, *Annu. Rev. Biochem.,* 52, 587, 1982.
90. **Dencher, N.,** Functions of bacteriorhodopsin, in *Biochemistry of Sensory Functions,* Jaenicke, L., Ed., Springer-Verlag, Berlin, 1974, 161.
91. **Schimz, A., Sperling, W., Ermann, P., Bestmann, H. J., and Hildebrand, E.,** Substitution of retinal by analogues in retinal pigments of *Halobacterium halobium.* Contributions of bacteriorhodopsin and halorhodopsin to photosensory activity, *Photochem. Photobiol.,* 38, 417, 1983.
92. **Spudich, J. L. and Stoeckenius, W.,** Photosensory and chemosensory behavior of *Halobacterium halobium, Photobiochem. Photobiophys.,* 1, 43, 1979.
93. **Dencher, N. A.,** Light-induced behavioural reactions of *Halobacterium halobium:* evidence for two rhodopsins acting as photopigments, in *Energetics and Microorganisms,* Caplan S. R. and Ginzburg, M., Eds., Elsevier/North-Holland, Amsterdam, Biomedical Press, 1978, 67.
94. **Dencher, N. A.,** Sensory transduction in *Halobacterium halobium:* retinal protein pigment controls UV-induced behavioral response, *Z. Naturforsch.,* 34c, 841, 1979.
95. **Sperling, W. and Schmiz, A.,** Photosensory retinal pigments in *Halobacterium halobium, Biophys. Struct. Mech.,* 6, 165, 1980.
96. **Hildebrand, E.,** Comparative discussion of photoreception in lower and higher organisms. Structural and functional aspects, in *Photoreception and Sensory Transduction in Aneural Organisms,* Lenci, F. and Colombetti, G., Eds., Plenum Press, New York, 1980, 319.
97. **Dencher, N. A. and Hildebrand, E.,** Photobehavior of *Halobacterium halobium, Meth. Enzymol.,* 88, 421, 1982.
98. **Schimz, A. and Hildebrand, E.,** *Halobacterium halobium,* mutants that are defective in photosensory and chemosensory behavior, *Hoppe-Seyler's Z. Physiol. Chem.,* 360, 1190, 1979.
99. **Takahashi, T. and Kobatake, Y.,** Computer-linked automated method for measurement of the reversal frequency in phototaxis of *Halobacterium halobium, Cell Struct. Func.,* 7, 183, 1982.
100. **Baryshev, V. A., Glagolev, A. N., and Skulachev, V. P.,** Sensing of $\Delta\psi_{H+}$ in phototaxis of *Halobacterium halobium, Nature (London),* 292, 338, 1981.
101. **Hildebrand, E. and Schmiz, A.,** Photosensory behavior of a bacteriorhodopsin-deficient mutant, ET-15, of *Halobacterium halobium, Photochem. Photobiol.,* 37, 581, 1983.

102. **Hazemoto, N., Kamo, N., Terayama, Y., Kobatake, Y., and Tsuda, M.,** Photochemistry of two rhodopsinlike pigments in bacteriorhodopsin-free mutant of *Halobacterium halobium, Biophys. J.,* 44, 59, 1983.

103. **Falke, J. J., Chan, S. I., Steiner, M., Oesterhelt, D., Towner, P., and Lanyi, J. K.,** Halide binding by the purified halorhodopsin chromoprotein. II. New chloride-binding sites revealed by ^{35}Cl NMR, *J. Histochem. Cytol.,* 259, 2185, 1984.

104. **Spudich, J. L. and Bogomolni, R. A.,** Spectroscopic discrimination of the three rhodopsinlike pigments in *Halobacterium halobium* membranes, *Biophys. J.,* 43, 243, 1983.

105. **Schimz, A., Sperling, W., and Hildebrand, E.,** Bacteriorhodopsin and the sensory pigment of the photosystem 565 in *Halobacterium halobium, Photochem. Photobiol.,* 36, 193, 1982.

106. **Traulich, B., Hildebrand, E., Schmiz, A., Wagner, G., and Lanyi, K.,** Halorhodopsin and photosensory behavior in *Halobacterium halobium* mutant strain L-33, *Photochem. Photobiol.,* 37, 577, 1983.

107. **Hildebrand, E. and Schmiz, A.,** Consecutive formation of sensory photosystems in growing *Halobacterium halobium, Photochem. Photobiol.,* 38, 593, 1983.

108. **Armitage, J. P. and Evans, M. C. W.,** The reaction in the phototactic and chemotactic response of photosynthetic bacteria, *FEMS Microbiol. Lett.,* 11, 89, 1981.

109. **Taylor, B. L., Miller, J. B., Warrick, H. M., and Koshland, D. E., Jr.,** Electron acceptor taxis and blue light effect on bacterial chemotaxis, *J. Bacteriol.,* 140, 567, 1979.

110. **Bovee, E. C.,** Movement and locomotion of *Euglena*, in *The Biology of Euglena,* Vol. 3, *Physiology,* Buetow, D. E., Ed., Academic Press, New York, 1982, 143.

111. **Colombetti, G., Lenci, F. and Diehn, B.,** Responses to photic, chemical, and mechanical stimuli, in *The Biology of Euglena,* Vol. 3, *Physiology,* Buetow, D. E., Eds., Academic Press, New York, 1982, 169.

112. **Morel-Laurens, N. M. L. and Feinleib, M. E.,** Photomovement in an "eyeless" mutant of *Chlamydomonas., Photochem. Photobiol.,* 37, 189, 1983.

113. **Melkonian, M. and Robenek, H.,** The eyespot of the flagellate *Tetraselmis cordiformis* Stein (Chlorophyceae): structural specialization of the outer chloroplast membrane and its possible significance in phototaxis of green algae, *Protoplasma,* 100, 183, 1979.

114. **Melkonian, M. and Robenek, H.,** Eyespot membranes of *Chlamydomonas reinhardii:* a freeze-fracture study, *J. Ultrastruct. Res.,* 72, 90, 1980.

115. **Melkonian, M. and Robenek, H.,** Eyespot membranes in newly released zoospores of the green alga *Chlorosarcinopsis gelatinosa* (Chlorosarcinales), and their fate during zoospore settlement, *Protoplasma,* 104, 129, 1980.

116. **Robenek, H. and Melkonian, M.,** Comparative ultrastructure of eyespot membranes in gametes and zoospores of the green alga *Ulva lactuca* (Ulvales), *J. Cell Sci.,* 50, 149, 1981.

117. **Risotori, T., Ascoli, C., Banchetti, R., Parrini, P., and Petracchi, D.,** Localization of photoreceptor and active membrane in the green alga *Haematococcus pluvialis,* in Progress in Protozoology, Abstr., 6th Int. Cong. Protozoology, Warsaw, July 5 to 11, 1981, 314.

118. **Song, P.-S.,** Protozoan and related photoreceptors: molecular aspects, *Annu. Rev. Biophys. Bioeng.,* 12, 35, 1983.

119. **Song, P.-S., Walker, E. B., Jung, J., Auerbach, R. A., Robinson, G. W., and Prezelin, B.,** Primary processes of photobiological receptors, in *New Horizons in Biological Chemistry,* Koike, M., et al., Eds., Academic Press, Tokyo, 1980, 79.

120. **Senger, H.,** The effect of blue light on plants and microorganisms, *Photochem. Photobiol.,* 35, 911, 1982.

121. **Raven, J. A.,** Do plant photoreceptors act at the membrane level?, *Phil. Trans. R. Soc. London Set. B,* 303, 403, 1983.

122. **Haupt, W. and Trump, K.,** Lichtorientierte Chloroplastenbewegung bei *Mougeotia*: die Grösse des Phytochromgradienten steuert die Bewegungsgeschwindigkeit, *Biochem. Physiol. Pflanz.,* 168, 131, 1975.

123. **Britz, S. J.,** Chloroplast and nuclear migration, in *Encyclopedia of Plant Physiology,* New Series, Vol. 7, *Physiology of Movements,* Haupt, W. and Feinleib, M. E., Eds., Springer-Verlag, Berlin, 1979, 170.

124. **Inoue, S.,** Video image processing greatly enhances contrast, quality and speed in polarization-based microscopy, *J. Cell Biol.,* 89, 346, 1981.

125. **Allen, R. D. and Allen, S.,** Videomicroscopy in the study of protoplasmic streaming and cell movement, *Protoplasma,* 109, 209, 1981.

126. **Lipson, E. D. and Häder, D.-P.,** Video data acquisition for movement responses in individual organisms, *Photochem. Photobiol.,* 39, 437, 1984.

127. **Häder, D.-P.,** Computer-based evaluation of phototactic orientation in microorganisms, *EDV Med. Biol.,* 12, 27, 1981.

128. **Berns, G. S. and Berns, M. W.,** Computer-based tracking of living cells, *Exp. Cell Res.,* 142, 103, 1982.

129. **Glazzard, A. N., Hirons, M. R., Mellor, J. S., and Holwill, M. E. J.,** The computer assisted analysis of television images as applied to the study of cell motility, *J. Submicrosc. Cytol.,* 15, 305, 1983.

130. **Mineyuki, Y., Yamada, M., Takagi, M., Wada, M., and Furuya, M.**, A digital image processing technique for the analysis of particle movements: its application to organelle movements during mitosis in *Adiantum protonemata, Plant Cell Physiol.*, 24, 225, 1983.
131. **Wilson, M. C. and Harvey, J. D.**, Twin-beam laser velocimeter for the investigation of spermatozoon motility, *Biophys. J.*, 41, 13, 1983.
132. **Racey, T. J. and Hallett, F. R.**, A quasi-elastic light scattering and cinematographical comparison of three strains of motile *Chlamydomonas reinhardtii:* a wild type strain, a colchicine resistant mutant and a backward swimming mutant, *J. Muscle Res. Cell Motil.*, 4, 333, 1983.
133. **Pfau, J., Nultsch, W., and Rüffer, U.**, A fully automated and computerized system for simultaneous measurements of motility and phototaxis in *Chlamydomonas, Arch. Microbiol.*, 135, 259, 1983.
134. **Feinleib, M. E.**, Behaviour models in flagellates, in *Biophysics of Photoreceptors and Photobehaviour of Microorganism*, Columbetti, G., Eds., Lito Felici, Pisa, 1975, 68.
135. **Boscov, J. S.**, Responses of *Chlamydomonas* to single flashes of light, thesis, Tufts University, Boston, 1974.
136. **Bendix, S. W.**, Phototaxis, *Bot. Rev.*, 26, 145, 1960.
137. **Häder, D.-P. and Wenderoth, K.**, Role of three basic light reactions in photomovement of desmids, *Planta*, 137, 207, 1977.
138. **Tevini, M. and Häder, D.-P.**, *Allgemeine Photobiologie*, Thieme, Stuttgart, 1985.
139. **Häder, D.-P., Claviez, M., Merkl, R., and Gerisch, G.**, Responses of *Dictyostelium discoideum* amoebae to local stimulation by light, *Cell Biol. Int. Rep.*, 7, 611, 1983.
140. **Grebecki, A.**, Effects of localized photic stimulation on amoeboid movement and their theoretical implications, *Eur. J. Cell Biol.*, 24, 163, 1981.
141. **Cullen, K. J. and Allen, R. D.**, A laser microbeam study of amoeboid movement, *Exp. Cell Res.*, 128, 353, 1980.
142. **Grebecki, A. and Klopocka, W.**, Functional interdependence of pseudopodia in *Amoeba proteus* stimulated by light-shade difference, *J. Cell Sci.*, 50, 245, 1981.
143. **Poff, K. L.**, Perception of a unilateral light stimulus, *Phil. Trans. Roy. Soc. London, Ser. B*, 303, 470, 1983.
144. **Häder, D.-P. and Burkart, U.**, Optical properties of *Dictyostelium discoideum* pseudoplasmodia responsible for phototactic orientation, *Exp. Mycol.*, 7, 1, 1983.
145. **Häder, D.-P.**, Negative phototaxis of *Dictyostelium discoideum* pseudoplasmodia in UV-radiation, *Photochem. Photobiol.*, 41, 225, 1985.
146. **Uematsu-Kaneda, H. and Furuya, M.**, Effects of viscosity on phototactic movement and period of cell rotation in *Cryptomonas sp., Physiol. Plant.*, 56, 194, 1982.
146a. **Häder, D.-P. and Lebert H.**, unpublished data, 1984.
147. **Satir, P.**, Mechanisms and controls of prokaryotic and eukaryotic flagellar motility, *Cell Biol. Int. Rep.*, 3, 641, 1979.
148. **Nultsch, W.**, The photocontrol of movement of *Chlamydomonas*, in *The Biology of Photoreception*, Cosens, D. J. and Vince-Price, D., Eds., Society for Experimental Biology Symposium No. 36, Cambridge University Press, Cambridge, 1983, 521.
149. **Hirschberg, R. and Hutchinson, W.**, Photoresponses of wild-type and mutant dikaryons of *Chlamydomonas, Curr. Microbiol.*, 4, 287, 1980.
150. **Goodenough, J. E., Bruce, V. G., and Cater, A.**, The effects of inhibitors affecting protein synthesis and membrane activity on the *Chlamydomonas reinhardii* phototactic rhythm, *Biol. Bull.*, 161, 371, 1981.
151. **Mikolajczyk, E. and Pado, R.**, The kinetics of photobehavior modification upon change of the suspension medium of *Euglena gracilis*, in *Progress in Protozoology*, Abstr. 6th Int. Congr. Protozoology, Warsaw, July 5 to 11, 1981, 253.
152. **Mikolajczyk, E. and Diehn, B.**, Morphological alteration in *Euglena gracilis* induced by treatment with CTAB (Cetyltrimethylammonium bromide) and Triton X-100: correlations with effects on photophobic behavioral responses, *J. Protozool.*, 25, 461, 1978.
153. **Goodenough, J. E. and Bruce, V. G.**, The effects of caffeine and theophylline on the phototactic rhythm of *Chlamydomonas reinhardii, Biol. Bull.*, 159, 649, 1980.
154. **Marchese-Ragona, S. P., Mellor, J. S., and Holwill, M. E. J.**, Calmodulin inhibitors cause flagellar wave reversal, *J. Submicrosc. Cytol.*, 15, 43, 1983.
155. **Hirschberg, R. and Hutchinson, W.**, Effect of chlorpromazine on phototactic behavior in *Chlamydomonas, Can. J. Microbiol.*, 26, 265, 1980.
156. **Simons, P. J.**, The role of electricity in plant movements, *New Phytol.*, 87, 11, 1981.
157. **Dolowy, K.**, Electrochemical model of cell membrane, cell adhesion and motility, *Stud. Biophys.*, 90, 185, 1982.
158. **Iwatsuki, K. and Song, P.-S.**, Heavy water (D_2O) enhances the photosensitivity of *Stentor coeruleus, Nature*, in press.

159. **Walker, E. B., Yoon, M., and Song, P.-S.,** The pH dependence of photosensory responses in *Stentor coeruleus* and model system, *Biochim. Biophys. Acta,* 634, 289, 1981.
160. **Baryshev, V. A., Glagolev, N. A., and Skulachev, V. P.,** The interrelation of phototaxis, membrane potential and K^+/Na^+ gradient in *Halobacterium halobium, J. Gen. Microbiol.,* 129, 367, 1983.
161. **Arshavsky, V. Y., Baryshev, V. A., Brown, I. I., Glagolev, A. N., and Skulachev, V. P.,** Transmembrane gradient of K^+ and Na^+ ions as an energy buffer in *Halobacterium halobium* cells, *FEBS Lett.,* 133, 22, 1981.
162. **Sakaguchi, H.,** Effect of external ionic environment on phototaxis of *Volvox carteri, Plant Cell Physiol.,* 20, 1643, 1979.
163. **Litvin, F. F., Sineshchekov, O. A., and Sineshchekov, V. A.,** Photoreceptor electric potential in the phototaxis of the alga *Haematococcus pluvialis, Nature (London),* 271, 476, 1978.
164. **Hyams, J. S. and Borisy, G. G.,** Isolated flagellar apparatus of *Chlamydomonas:* characterization of forward swimming and alteration of waveform and reversal of motion by calcium ions *in vitro, J. Cell. Sci.,* 33, 235, 1978.
165. **Marme, D. and Dieter, P.,** Role of Ca^{2+} and calmodulin in plants, in *Calcium and Cell Function,* Vol. 4, Cheung W. Y., Eds., Academic Press, New York, 1983, 264.
166. **Uematsu-Kaneda, H. and Furuya, M.,** Effects of calcium and potassium ions on phototaxis in *Cryptomonas, Plant Cell Physiol.,* 23, 1377, 1982.
167. **Baryshev, V. A., Glagolev, A. N., and Skulachev, V. P.,** Interrelationship between Ca^{2+} and a methionine-requiring step in *Halobacterium halobium* taxis, *FEMS Microbiol. Lett.,* 13, 47, 1981.
168. **Baryshev, V. A.,** Regulation of *Halobacterium halobium* motility by Mg^{2+} and Ca^{2+} ions, *FEMS Microbiol. Lett.,* 14, 139, 1982.
169. **Murvanidze, G. V. and Glagolev, A. N.,** Calcium ions regulate reverse motion in phototactically active *Phormidium uncinatum,* and *Halobacterium halobium, FEMS Microbiol. Lett.,* 12, 3, 1981.
170. **Dohrmann, U., Fisher, P. R., Brüderlein, M., and Williams, K. L.,** Transitions in *Dictyostelium discoideum* behaviour: influence of calcium and fluoride on slug phototaxis and thermotaxis, *J. Cell Sci.,* 65, 111, 1984.
171. **Prusti, R. K., Song, P.-S., Häder, D.-P., and Häder, M.,** Caffeine-enhanced photomovement in the ciliate *Stentor coeruleus, Photochem. Photobiol.,* 40, 369, 1984.
172. **Watanabe, M. and Furuya, M.,** Phototactic behavior of individual cells of *Cryptomonas* sp. in response to continuous and intermittent light stimuli, *Photochem. Photobiol.,* 35, 559, 1982.
173. **Salisbury, J. L.,** Contractile flagellar roots: the role of calcium, *J. Submicrosc. Cytol.,* 15, 105, 1983.
174. **Smyth, R. D. and Berg, H. C.,** Change in flagellar beat frequency of *Chlamydomonas* in response to light, *Cell Motil. Suppl.,* 1, 211, 1982.
175. **Kamiya, R. and Witman, G. B.,** Submicromolar levels of calcium control the balance of beating between the two flagella in demembranated models of *Chlamydomonas, J. Cell Biol.,* 98, 97, 1984.
176. **Doughty, M. J., Grieser, R., and Diehn, B.,** Photosensory transduction in the flagellated alga *Euglena gracilis, Biochim. Biophys. Acta,* 602, 10, 1980.
177. **Doughty, M. J. and Diehn, B.,** Photosensory transduction in the flagellated alga, *Euglena gracilis.* I. Action of divalent cations, Ca^{2+} antagonists and Ca^{2+} ionophore on motility and photobehavior, *Biochim. Biophys. Acta,* 588, 148, 1979.
178. **Doughty, M. J. and Diehn, B.,** Photosensory transduction in the flagellated alga, *Euglena gracilis.* III. Induction of Ca^{2+}-dependent responses by monovalent cation ionophores, *Biochim. Biophys. Acta,* 682, 32, 1982.
179. **Doughty, M. J. and Diehn, B.,** Photosensory transduction in the flagellated alga, *Euglena gracilis.* IV. Long term effects of ions and pH on the expression of step-down photobehaviour, *Arch. Microbiol.,* 134, 204, 1983.
180. **Bibikov, S. I., Baryshev, V. A., and Glagolev, A. N.,** The role of methylation in the taxis of *Halobacterium halobium* to light and chemo-effectors, *FEBS Lett.,* 146, 255, 1982.
181. **Schimz, A.,** Methylation of membrane proteins is involved in chemosensory and photosensory behavior of *Halobacterium halobium, FEBS Lett.,* 125, 205, 1981.
182. **Schimz, A.,** Localization of the methylation system involved in sensory behavior of *Halobacterium* and its dependence on calcium, *FEBS Lett.,* 139, 283, 1982.
183. **Spudich, E. N. and Spudich, J. L.,** Measurement of light-regulated phosphoproteins of *Halobacterium halobium, Meth. Enzymol.,* 88, 213, 1982.
184. **Akitaya, T., Hirose, T., Ueda, T., and Kobatake, Y.,** Variation of intracellular cyclic AMP and cyclic GMP following chemical stimulation in relation to contractility in *Physarum polycephalum, J. Gen. Microbiol.,* 130, 549, 1984.
185. **Brenner, M. and Thoms, S. D.,** Caffeine blocks activation of cyclic AMP synthesis in *Dictyostelium discoideum, Dev. Biol.,* 101, 136, 1984.

186. **Melkonian, M.,** The functional analysis of the flagellar apparatus in green algae, in *Prokaryotic and Eukaryotic Flagella,* Amos, W. B. and Duckett, J. G., Eds., Society for Experimental Biology Symposium No. 35, Cambridge University Press, Cambridge, 1982, 589.

187. **Sleigh, M. A.,** Flagellar beat patterns and their possible evolution, *BioSystems,* 14, 423, 1981.

188. **Hines, M. and Blum, J. J.,** Three-dimensional mechanics of eukaryotic flagella, *Biophys. J.,* 41, 67, 1983.

189. **Kamiya, R., Nagai, R., and Nakamura, S.,** Rotation of the central-pair microtubules in *Chlamydomonas* flagella, in *Biological Functions of Microtubules and Related Structures,* Sakai, H., Mohri, H., and Borisy, G. G., Eds., Academic Press, Tokyo, 1982, 189.

190. **Herth, W.,** Twist (and rotation?) of central-pair microtubules in flagella of *Poterioochromonas, Protoplasma,* 112, 17, 1982.

191. **Hoops, H. J. and Witman, G. B.,** Outer doublet heterogeneity reveals structural polarity related to beat direction in *Chlamydomonas flagella, J. Cell Biol.,* 97, 902, 1983.

192. **Kivic, P. A. and Walne, P. L.,** Algal photosensory apparatus probably represent multiple parallel evolutions, *BioSystems,* 16, 31, 1983.

193. **Carlile, M. J.,** The biological significance and evolution of photosensory system, in *The Blue Light Syndrome,* Senger, H., Ed., Springer Verlag, Berlin, 1980, 2.

194. **Carlile, M. J.,** The photobiology of fungi, *Annu. Rev. Plant Physiol.,* 16, 175, 1965.

195. **Haupt, W.,** Orientierungsbewegungen bei niederen Pflanzen und Bakterien. Erkenntnisse der biologischen Forschung, *Universitas, Ztschr. Wiss. Kunst, Lit.,* 36, 531, 1981.

196. **Haupt, W.,** Physiology of movement, *Prog. Bot.,* 44, 222, 1982.

197. **Haupt, W.,** Localization and orientation of photoreceptor pigments, in *Photoreception and Sensory Transduction in Aneural Organisms,* Lenci, F. and Colombetti, G., Eds., Plenum Press, New York, 1980, 155.

198. **Alldredge, A. L. and King, J. M.,** Effects of moonlight on the vertical migration patterns of demersal zooplankton, *J. Exp. Mar. Biol. Ecol.,* 44, 133, 1980.

199. **Häder, D.-P.,** Effects of UV-B on motility and photorientation in the cyanobacterium, *Phormidium uncinatum, Arch. Microbiol.,* 140, 34, 1984.

200. **Burkart, U. and Häder, D.-P.,** Phototactic attraction in light trap experiments: a mathematical model, *J. Math. Biol.,* 10, 257, 1980.

201. **Pentecost, A.,** Effects of sedimentation and light intensity on matforming Oscillatoriaceae with particular reference to *Microcoleus lyngbyaceus* Gomont, *J. Gen. Microbiol.,* 130, 983, 1984.

202. **Häder, D.-P.,** Wie orientieren sich Cyanobakterien im Licht, *BIUZ,* 14, 78, 1984.

203. **Fay, P.,** *The Blue-Greens,* The Institute of Biology's Studies in Biology, No. 160, Edward Arnold, London, 1983.

204. **Castenholz, R. W.,** Motility and taxes, in *The Biology of Cyanobacteria, Botanical Monographs,* Vol. 19, Carr, N. G. and Whitton, B. A., Eds., Blackwell Scientific Publ., Oxford, 1982, 413.

205. **Nelson, D. C. and Castenholz, R. W.,** Light responses of *Beggiatoa, Arch. Microbiol.,* 131, 146, 1982.

206. **Nultsch, W. and Häder, D.-P.,** Bestimmungen der photophobotaktischen Unterschiedsschwelle bei *Phormidium uncinatum, Ber. Dtsch. Bot. Ges.,* 83, 185, 1970.

207. **Häder, D.-P.,** Inhibition of phototaxis and motility by UV-B irradiation in *Dictyostelium discoideum* slugs, *Plant Cell Physiol.,* 24, 1545, 1983.

208. **Häder, D.-P.,** Effects of UV-B on motility and photobehavior in the flagellate, *Euglena gracilis, Arch. Microbiol.,* 141, 159, 1985.

209. **Häder, D.-P.,** Effects of UV-B irradiation on sorocarp development of *Dictyostelium discoideum, Photochem. Photobiol.,* 38, 551, 1983.

210. **Maugh, T. H., II,** What is the risk from chlorofluorocarbons?, *Science,* 223, 1051, 1984.

211. **Seitz, K.,** Cytoplasmic streaming and cyclosis of chloroplasts, in *Encylopedia of Plant Physiology,* N.S. Vol. 7, *Physiology of Movements,* Haupt, W. and Feinleib, M. E., Eds., Springer-Verlag, Berlin, 1979, 150.

212. **Takagi, S. and Nagai, R.,** Regulation of cytoplasmic streaming in *Vallisneria* mesophyll cells, *J. Cell Sci.,* 62, 385, 1983.

213. **Seitz, K.,** Light dependent control of cytoplasmic streaming and chloroplast movement, *Acta Protozool.,* 18, 197, 1978.

214. **Filner, P. and Yadav, B. S.,** Intracellular movements. Role of microtubules in intracellular movements, in *Encyclopedia of Plant Physiology,* N.S. Vol. 7, *Physiology of Movements,* Haupt, W. and Feinleib, M. E., Eds., Springer-Verlag, Berlin, 1979, 95.

215. **Haupt, W.,** Light-mediated movement of chloroplasts, *Annu. Rev. Plant Physiol.,* 33, 205, 1982.

216. **Seitz, K.,** Chloroplast motion in response to light in aquatic vascular plants, in *Studies on Aquatic Vascular Plants,* Symoens, J. J., Hooper, S. S., and Compere, P., Eds., Royal Botanical Society, Brussels, 1982, 89.

217. **Nothnagel, E. A. and Webb, W. W.,** Hydrodynamic models of viscous coupling between motile myosin and endoplasm in Characean algae, *J. Cell Biol.,* 94, 444, 1982.

218. **Seitz, K.,** Light induced changes in the centrifugability of chloroplasts mediated by an irradiance dependent interaction of respiratory and photosynthetic processes, in *The Blue Light Syndrome,* Senger, H., Eds., Springer-Verlag, Berlin, 1980, 637.

219. **Kurdoa, K.,** Cytoplasmic streaming in Characean cells cut open by microsurgery, *Proc. Jpn. Acad.,* 59, 126, 1983.

220. **Williamson, R. E. and Ashley, C. C.,** Free Ca^{2+} and cytoplasmic streaming in the alga *Chara, Nature,* 296, 647, 1982.

221. **Kuroda, K. and Kamiya, N.,** Active movement of *Nitella* chloroplasts *in vitro, Proc. Jpn. Acad.,* 51, 774, 1975.

222. **Kikuyama, M. and Tazawa, M.,** Ca^{2+} ion reversibly inhibits the cytoplasmic streaming of *Nitella, Protoplasma,* 113, 241, 1982.

223. **Chen, J. C. W.,** Effects of elevated centrifugal field on the *Nitella* cell and postcentrifugation patterns of its cytoplasmic streaming and chloroplast files, *Cell Struct. Func.,* 8, 109, 1983.

224. **Sheetz, M. P. and Spudich, J. A.,** Movement of myosin-coated structures on actin cables, *Cell Motil.,* 3, 485, 1983.

225. **Kuroda, K. and Manabe, E.,** Microtubule-associated cytoplasmic streaming in *Caulerpa, Proc. Jpn. Acad.,* 59, 131, 1983.

226. **Shimmen, T. and Tazawa, M.,** Control of cytoplasmic streaming by ATP, Mg^{2+} and cytochalasin B in permeabilized Characeae cell, *Protoplasma,* 115, 18, 1983.

227. **Mizukami, M. and Wada, S.,** Action spectrum for light-induced chloroplast accumulation in a marine coenocytic green alga, *Bryopsis plumosa, Plant Cell Physiol.,* 22, 1245, 1981.

228. **Blatt, M. R. and Briggs, W. R.,** Quantitative microphotometry at the cellular level: a simple technique for measuring chloroplast movements *in vivo, Photochem. Photobiol.,* 38, 347, 1983.

229. **Blatt, M. R.,** The action spectrum for chloroplast movements and evidence for blue-light-photoreceptor cycling in the alga *Vaucheria, Planta,* 159, 267, 1983.

230. **Blatt, M. R., Weisenseel, M. H., and Haupt, W.,** A light-dependent current associated with chloroplast aggregation in the alga *Vaucheria sessilis, Planta,* 152, 513, 1981.

231. **Haupt, W.,** The perception in light direction and orientation responses in chloroplasts, in *The Biology of Photoreception,* Cosens, D. J. and Vince-Price, D., Eds., Society for Experimenta Biology, Cambridge, 1983, 423.

232. **Haupt, W.,** Wavelength-dependent action dichroism: a theoretical consideration, *Photochem. Photobiol.,* 39, 107, 1984.

233. **Haupt, W.,** Movement of chloroplasts under the control of light, in *Progress in Phycological Research,* Vol. 2, Round, F. E. and Chapman, D. J., Eds., Elsevier, Amsterdam, 1983, 227.

234. **Rüffer, U., Pfau, J., and Nultsch, W.,** Movements and arrangements of *Dictyota* phaeoplasts in response to light and darkness, *Z. Pflanzphysiol.,* 101, 283, 1981.

235. **Nultsch, W., Pfau, J., and Rüffer, U.,** Do correlations exist between chromatophore arrangement and photosynthetic activity in seaweeds?, *Mar. Biol.,* 62, 111, 1981.

236. **Zurzycki, J., Walczak, T., Gabrys, H., and Kajfosz, J.,** Chloroplast translocations in *Lemna trisulca* L. induced by continuous irradiation and by light pulses, *Planta,* 157, 502, 1983.

237. **Gabrys, H., Walczak, F., and Zurzycki, J.,** Chloroplast translocation induced by light pulses. Effects of single light pulses, *Planta,* 152, 553, 1981.

238. **Schmid, R.,** Effects of blue light on the circadian rhythm in *Acetabularia mediterranea, Protoplasma,* 105, 364, 1980.

239. **Schmid, R. and Koop, H.-U.,** Chloroplast banding during the circadian chloroplast migration in *Acetabularia, Protoplasma,* 105, 364, 1980.

240. **Paques, M. and Brouers, M.,** Chloroplast phototaxis in *Acetabularia mediterranea, Protoplasma,* 105, 360, 1980.

241. **Schmid, R. and Koop, H.-U.,** Properties of the chloroplast movement during the circadian chloroplast migration in *Acetabularia mediterranea, Z. Pflanzenphysiol.,* 112, 351, 1983.

242. **Britz, S. J. and Briggs, W. R.,** Rhythmic chloroplast migration in the green alga *Ulva:* dissection of movement mechanism by differential inhibitor effects, *Eur. J. Cell Biol.,* 31, 1, 1983.

243. **Töpperwien, F. and Hardeland, R.,** Free-running circadian rhythm of plastid movement in individual cells of *Pyrocystis lunula* (Dinophyta), *J. Interdiscip. Cycle Res.,* 11, 325, 1980.

244. **Schönbohm, E., Schönbohm, E. and Lücke, G.,** Die Entstehung symmetrischer und asymmetrischer Phytochrom-Gradienten bei *Mougeotia* und deren Bedeutung für die Chloroplastenorientierung im Stark- und Schwachlicht, *Ber. Dtsch. Bot. Ges.,* 92, 297, 1979.

245. **Haupt, W. and Wachter, W.,** Steuerung der Chloroplastenbewegung von *Mougeotia* durch Absorptions- gradienten in Rot- und Blaulicht, *Z. Pflanzenphysiol.,* 96, 211, 1980.

246. **Kraml, M. and Haupt, W.,** Phytochrome-controlled chloroplast orientation in *Mougeotia:* action dichroism in double-flash experiments, *Plant Sci. Lett.,* 21, 145, 1981.

247. **Wagner, G. and Klein, K.,** Mechanism of chloroplast movement in *Mougeotia, Protoplasma,* 109, 169, 1981.

248. **Schönbohm, E.,** Phytochrome and non-phytochrome dependent blue light effects on intracellular movements in fresh-water algae, in *The Blue Light Syndrome,* Senger, H., Ed., Springer-Verlag, Berlin, 1980, 69.

249. **Schönbohm, E. and Schönbohm, E.,** Zur Problematik des *In-vivo-* Einsatzes von Kaliumjodid und Natriumazid als spezifische Triplett-Quencher von Flavinen, *Biochem. Physiol. Pflanz.,* 179, 95, 1984.

250. **Kraml, M.,** Photoconversion of phytochrome by short red flashes in *Mougeotia* and *Avena,* in *Photoreceptors and Plant Development,* de Greef, J., Ed., Antwerp University Press, Antwerp, 1960, 59.

251. **Haupt, W. and Polacco, E.,** Phytochrome-mediated response in *Mougeotia* to very short laser flashes, *Plant Sci. Lett.,* 17, 67, 1979.

252. **Haupt, W., Hupfer, B., and Kraml, M.,** Blitzlichtinduktion der Chloroplastenbewegung bei *Mougeotia:* Wirkung unterschiedlicher Spektralbereiche und Polarisationsrichtungen, *Z. Pflanzenphysiol.,* 96, 331, 1980.

253. **Haupt, W. and Reif, G.,** "Ageing" of phytochrome P_{fr} in *Mesotaenium, Z. Pflanzenphysiol.,* 92, 153, 1979.

254. **Schönbohm, E. and Brühl, K.-L.,** Ein neuer Inversionstyp der Schwachlichtbewegung des *Mougeotia*-Chloroplasten im längsschwingenden polarisierten Light, *Ber. Dtsch. Bot. Ges.,* 92, 305, 1979.

255. **Schönbohm, E. and Schönbohm, E.,** Über die energetische Steuerung der Phytochrom-kontrollierten Chloroplastenverankerung im cytoplasmatischen Wandbelag, *Biochem. Physiol. Pflanz.,* 179, 473, 1984.

256. **Gabrys, H., Walczak, T., and Haupt, W.,** Blue-light-induced chloroplast orientation in *Mougeotia.* Evidence for a separate sensor pigment besides phytochrome, *Planta,* 160, 21, 1984.

257. **Wagner, G. and Rossbacher, R.,** X-Ray microanalysis and chlorotetracycline staining of calcium vesicles in the green alga *Mougeotia, Planta,* 149, 298, 1980.

258. **Schönbohm, E.,** Durch Phytochrom aktivierbare kontraktile Plasmaelemente: Lokalisation in der *Mougeotia-* Zelle. V. Mitteilung zur Mechanik der Chloroplastenbewegung, *Z. Pflanzenphysiol.,* 3, 185, 1979.

259. **Tendel, J. and Haupt, W.,** Mechanische und energetische Grundlagen der lichtabhängigen Gestaltänderung des *Mougeotia* chloroplasten, *Z. Pflanzenphysiol.,* 104, 169, 1981.

260. **Poff, K. L., Fontana, D. R., Häder, D.-P., and Schneider, M. J.,** A model for phototactic orientation by *Dictyostelium discoideum* slugs, *Plant Cell Phys.,* 27, 221, 1986.

Interaction with Other Light Effects

Chapter 10

MODE OF COACTION BETWEEN BLUE/UV LIGHT AND LIGHT ABSORBED BY PHYTOCHROME IN HIGHER PLANTS

H. Mohr

TABLE OF CONTENTS

I. SENSOR PIGMENTS IN HIGHER PLANTS

All life on earth is fueled by sunlight. In order to harvest the light quanta most successfully in the process of photosynthesis, plants must adapt to the light conditions of their particular habitat. In fact, development of photoautotrophic higher plants is "opportunistic" in the sense that the developmental process is in part controlled by light. It is only the basic developmental patterns of plant construction whch are *strictly* determined by the genes ("developmental homeostasis"); within these limits fine tuning of developmental events is controlled by the actual light climate at the site where the plant has to grow.[1]

In order to respond properly, a plant has to sense the light conditions in its environment continuously and accurately. This implies that a plant must be capable of sensing the quality and quantity of light throughout the spectrum of the sun as far as sunlight leads to electronic excitations (290 to 800 nm). For reasons of molecular physics it is improbable that a single photoreceptor can fulfill this task. Rather, we might expect that a higher plant will use several sensor pigments to check the whole spectral range with the sensitivity and accuracy required.

As far as we know today, three different sensor pigments occur in higher plants.[2] These are phytochrome (operating predominantly in the red/far-red spectral range), cryptochrome (operating in the blue/UV-A spectral range), and a UV-B photoreceptor. The action spectrum related to the latter photoreceptor shows a single intense peak at 290 nm and no action at wavelengths longer than 350 nm.[3] While some plants (e.g., Schirokko wheat[4]) seem to use the UV-B photoreceptor and phytochrome only, others (such as the milo seedling[5]) are obviously capable of sensing the light conditions throughout the sun's spectrum from UV-B to the near infrared.

II. PHYTOCHROME — A PHOTOCHROMIC PIGMENT

Phytochrome is the best-known sensor pigment in higher plants. Its existence and its photochromic properties were predicted by Hendricks and Borthwick in 1952 on the basis of purely physiological experiments.[6] Today we know that the phytochrome molecule is a soluble chromoprotein containing a linear tetrapyrrol chromophore (Figure 1) covalently linked to a polypeptide of 120 to 127 kilodaltons, depending on the plant species.[8] Phytochrome exists in two forms: P_r, which absorbs maximally in the red spectral range at around 665 nm; and P_{fr}, which absorbs maximally in the far-red spectral range at around 725 nm. These two forms are reversibly interconvertible by light. In a dark-grown plant, only P_r is present.

The photoconversion of P_r to P_{fr} in the cell induces a large number of diverse photoresponses. Reconversion of P_{fr} to P_r cancels these inductions. Thus, phytochrome is considered to function as a reversible biological switch, with P_{fr} as the physiologically active species.[9] Even though the molecular "mechanism" (i.e., the sequence of molecular steps) by which P_{fr} induces the observed developmental changes is still a matter of research and debate, it has now been verified that phytochrome-regulated development involves gene expression.[10,11] The regulation appears to occur at the transcriptional level.[12,13] Phytochrome-mediated changes in gene expression are measurable within 15 to 30 min of irradiation.

Both forms of phytochrome have broad overlapping absorption spectra throughout the visible range.[2] This is the reason why a photoequilibrium is achieved if visible light of a sufficiently high fluence rate is falling on a plant (or on a phytochrome preparation in the test tube). The wavelength-dependent photoequilibrium is usually defined by

$$\varphi_\lambda = \frac{[P_{fr}]_\lambda}{[P_{tot}]}$$

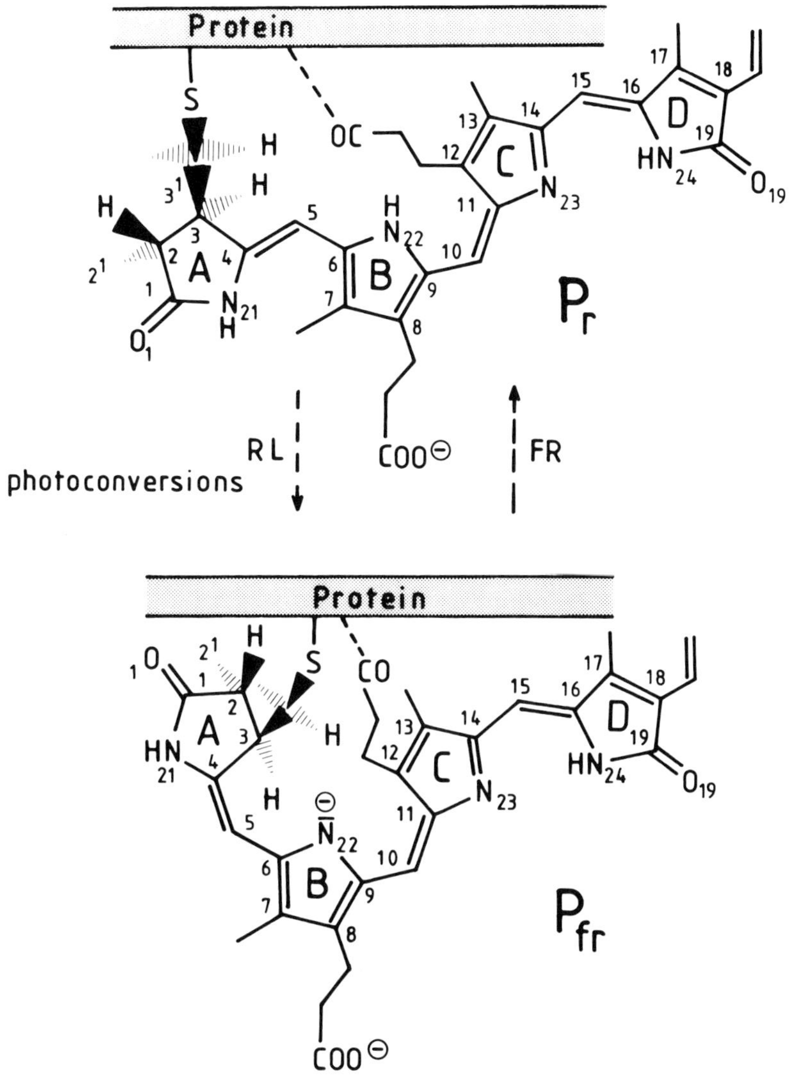

FIGURE 1. Suggested structures of the P_r and P_{fr} chromophores of phytochrome and their binding to the apoprotein (RL, red light; FR, far-red light).[7]

whereby P_{tot} (total phytochrome) is the sum of P_r and P_{fr}. Photoequilibrium is rapidly established in the red and far-red range of the spectrum since absorbance of both phytochrome forms is high in this range. As an example, at a fluence rate of 7 W · m^{-2} only "light pulses" (e.g., 1 min of light) are required to establish photoequilibria of the phytochrome system in a long-wavelength visible light.

The situation in short-wavelength visible light is not so unequivocal. This is partly due to the low absorbance of phytochrome in this spectral range and partly caused by increased scattering and light attenuation by carotenoid and flavin absorption which interfere with φ measurements. However, in a number of tissues (e.g., hypocotyls of dicotyledonous seedlings) $\varphi_{450\ nm}$ was found to be of the order of 0.4, whereas in the red light (RL) $\varphi_{660\ nm}$ is 0.8. The photoequilibrium (φ) established by medium far-red light (720 nm) is 0.03 and that established by long-wavelength far-red light (756 nm) is less than 0.01.[14]

Table 1
INDUCTION AND FULL REVERSION OF
THIS INDUCTION IN THE CASE OF
ANTHOCYANIN SYNTHESIS IN
SEEDLING COTYLEDONS OF MUSTARD
(*SINAPIS ALBA* L.)

Light treatment	Relative amount of anthocyanin (%)
5 min 660 nm	100
5 min 720 nm	49
5 min 660 nm + 5 min 720 nm	51
5 min 756 nm	32
5 min 660 nm + 5 min 756 nm	32
5 min 720 nm + 5 min 756 nm	35
Dark control	9

Note: Red ($\varphi_{660\,nm} = 0.8$), medium far-red ($\varphi_{720\,nm} = 0.03$) and long-wavelength far-red ($\varphi_{756\,nm} < 0.01$) light pulses were given to dark-grown seedlings at 36 hr after sowing. Anthocyanin was assayed 24 hr after the light pulse treatment. (Data of R. Schmidt.) If a dark interval is inserted between the two pulses, an escape from full reversibility is eventually observed. Percent reversibility is calculated according to the formula: % reversibility = $(a - c)/a - b) \cdot 100$, where a is the extent of the response obtained with a saturating RL pulse; b is the response obtained with a saturating far-red light pulse given at (approximately) the same time as the RL pulse; and c is the response obtained with the program: red light pulse—dark interval—far-red light pulse.

III. CRITERIA FOR PHYTOCHROME INVOLVEMENT IN RESPONSES CAUSED BY LIGHT PULSES

The criteria are defined as follows.[2] If induction of a photoresponse by a RL pulse ($\varphi_{RL} = 0.8$) is fully reversed by following it with a saturating pulse of far-red light ($\varphi_{720\,nm} = 0.03$, $\varphi_{756\,nm} < 0.01$) the response is controlled by phytochrome. These criteria have been verified in many instances, with a typical example being light-mediated anthocyanin synthesis in the mustard seedling (Table 1).

IV. MODE OF COACTION BETWEEN PHYTOCHROME AND THE PHOTORECEPTORS ABSORBING BLUE/UV LIGHT

There are (at least) two possibilities for the mode of coaction between phytochrome and the photoreceptors absorbing blue/UV light: independent and interdependent (Table 2).

A. Independent Coaction

The different sensor pigments operate independently of each other, eventually causing the same terminal photoresponse (e.g., anthocyanin synthesis). While an independent action would not necessarily lead to an *additive* coaction (one might expect a nonlinear relationship between *total* stimulus and response), any single photoreceptor must be expected to elicit a significant response.

Table 2
SUGGESTED MODES OF COACTION BETWEEN
PHYTOCHROME AND THE PHOTORECEPTORS
ABSORBING BLUE/UV LIGHT

Independent coaction

Phytochrome ⟶ Photoresponse *a*
Cryptochrome ⟶ Photoresponse *a*
UV-B photoreceptor ⟶ Photoresponse *a*

Interdependent coaction[a]

a The dashed line is to indicate that light absorbed by phytochrome
can strongly increase the effectiveness of P_{fr} in mediating a response. As an example, in the mustard seedling synthesis of
anthocyanin can be induced by single light pulses that operate
through phytochrome (see Table 1). However, the effectiveness
of phytochrome (P_{fr}) in mediating anthocyanin synthesis in the
epidermal cells of mustard cotyledons can be increased strongly
by a light pretreatment of the seedling.[9,15] The effective light is
absorbed by phytochrome. Thus, a light pretreatment, operating through phytochrome, leads to a strong, albeit transient,
amplification of responsivity in phytochrome (P_{fr})-mediated
anthocyanin synthesis of the mustard seedling.

B. Interdependent Coaction

The sensor pigments depend on each other to bring about the final photoresponse (e.g.,
anthocyanin synthesis). As an example, phytochrome (P_{fr}) may be the only effector molecule
to operate on gene expression in photomorphogenesis; however, to establish or to increase
and maintain responsivity in a plant cell towards P_{fr}, concomitant light absorption in cryptochrome or UV-B photoreceptor might be required. It seems in fact that a coaction of this
kind between phytochrome and the blue/UV light sensor pigments is characteristic for most
photomorphogenesis.[16,17] This coaction must be considered as highly economic since a single
"effector" (namely P_{fr}) suffices to bring about the molecular events leading to photomorphogenesis, and yet information about the whole solar spectrum — as far as it is relevant
for the plant — can contribute to the *rate and extent* of the photomorphogenetic responses.

The following representative case studies show that the kind of coaction as suggested by
Table 2 in fact operates in nature.

V. SELECTED CASE STUDIES

A. Anthocyanin Formation in the Milo Seedling (*Sorghum vulgare* Pers., cv. **Weider Hybrid**)

The mesocotyl of the milo seedling does not produce anthocyanin in complete darkness.
As described originally by Downs and Siegelman[18] even long-term red or far-red light does
not lead to any anthocyanin synthesis, while white light (WL) or blue light/ultraviolet (BL/
UV) cause strong and rapid pigmentation (Figure 2). Experiments with red and far-red light

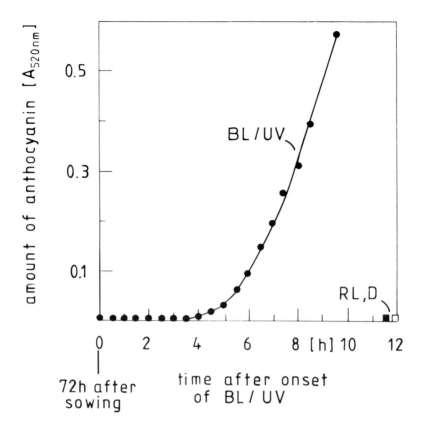

FIGURE 2.   Time course of anthocyanin accumulation in mesocotyl of the milo seedling in continuous blue/UV light (9.6 W · m^{-2}). The spectral radiant power distribution of the light source shows a peak in the UV-A region at 350 nm.[5,19] (Data of R. Oelmüller.)

pulses given after an inductive WL period of 3 hr have shown that phytochrome can act once a BL/UV effect has occurred.[20] On the other hand, the expression of the BL/UV effect is controlled by phytochrome. In experiments with dichromatic irradiations, i.e., simultaneous irradiation with two kinds of light to strongly modulate the level of P_{fr} on a constant background of BL/UV, its was found that the blue/UV light photoreaction as such is not affected by the presence or virtual absence of P_{fr}.[20]

The tentative interpretation of the data obtained with milo was that phytochrome (P_{fr}) is the effector which causes anthocyanin synthesis through activation of competent genes, while the BL/UV effect must be considered as establishing responsivity towards P_{fr}. In the case of anthocyanin synthesis in the milo mesocotyl, there is no responsivity without the operation of a BL/UV photoreceptor.

The action of BL/UV in anthocyanin synthesis of the milo mesocotyl shows the characteristics of an induction. As shown in Figure 3, even 5 min of BL/UV suffice to induce responsivity towards RL, i.e., towards P_{fr}. However, BL/UV is ineffective — as far as anthocyanin formation is concerned — without P_{fr}. This is documented by the ineffectiveness (with regard to appearance of anthocyanin) of the BL/UV treatment up to 10 min, provided that virtually all P_{fr} is returned to P_r at the end of the BL/UV treatment by a long-wavelength far-red light pulse ($\varphi_{RG9\ light} < 0.01$). Clearly, 5 min of BL/UV achieve something, namely, responsivity to P_{fr}. However, 5 min of BL/UV *alone* do not suffice to cause anthocyanin synthesis. On the other hand, BL/UV induces responsivity to P_{fr} so rapidly that after 15 min of BL/UV the light induction is no longer fully reversible. This means that in the presence

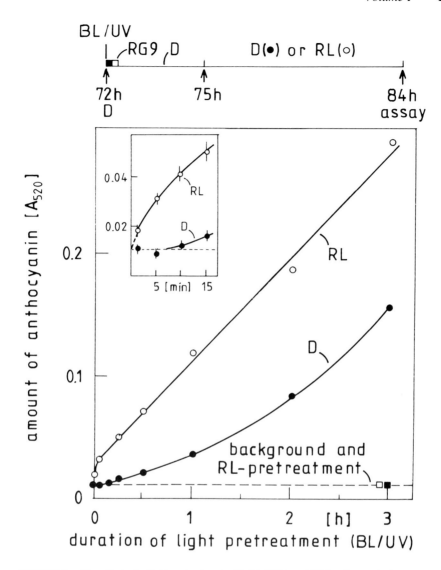

FIGURE 3. The effect of a light pretreatment with BL/UV (onset 72 hr after sowing) on phytochrome-mediated synthesis of anthocyanin in the mesocotyl of the milo seedling. The experimental protocol is indicated above. A saturating 5-min RG9 light pulse was given immediately after the end of the BL/UV pretreatment to return almost all P_{fr} back to P_r (φ_{RG9} < 0.01). The photoequilibrium in blue/UV light (see legend to Figure 2) is attained within 5 min, $\varphi_{BL/UV}$ = 0.74. The RL treatment (6.8 W · m^{-2}) was given continuously between 75 and 84 hr. Without the BL/UV treatment there is absolutely no response to RL treatment of any kind. D (●) means that the seedling received the pretreatment only but no RL between 75 and 84 hr. The inset enlarges the results obtained during the first 15 min of light pretreatment (different set of experiments).[27]

of BL/UV, P_{fr} can perform its initial action* within 15 min even though it requires 3 hr before anthocyanin appears (see Figure 2).

B. Hypocotyl Growth in Sesame (*Sesamum indicum* L.)

The rate of hyptocotyl elongation in the sesame seedling is controlled by WL and BL

---

* The term "initial action" designates the action of P_{fr} on some cell function which is no longer reversible by the removal of P_{fr}. The onset of the initial action is defined by the escape from full reversibility (see the footnote to Table 1).

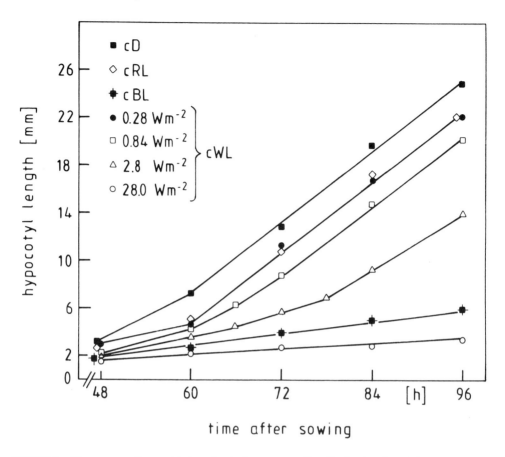

FIGURE 4.   Time courses of hypocotyl elongation in the sesame seedling. Explication of symbols: cD, continuous darkness; cRL, continuous low-fluence-rate RL (0.68 W · m^{-2}); cBL, continuous BL (7 W · m^{-2}); cWL, continuous WL of different energy fluence rates (W · m^{-2}). Light given from time of sowing.[21]

while RL — continuous as well as repeated pulses — is totally ineffective beyond 60 hr after sowing (Figure 4).[21]

Between 36 and 60 hr after sowing the growth rate responds strongly to RL pulses, the effect of which is fully reversible by far-red light.[21] Thus, an action of phytochrome is indicated, but only up to 60 hr. However, when seedlings are kept in BL for 3 days and then treated with RL, hypocotyl growth rate responds strongly.[21] These data and a large number of further experiments show that BL is required to establish and to maintain responsivity toward P_{fr}. As an example, Figure 5 shows that continuous BL inhibits hypocotyl growth with a lag of somewhat less than 2 hr (▲). A transfer after 2 hr of BL to darkness (△) leads to a return of the growth rate to the dark-specific growth rate within 12 hr. If most of P_{fr} is returned to P_r at the end of the 2-hr BL period by a 5-min RG 9 light pulse ($\varphi_{RG\ 9} < 0.01$), growth rate (□) returns to the growth rate in darkness (○) within 6 hr. If the seedlings are given a 5-min RL pulse before transfer to darkness ($\varphi_{RL} = 0.8$), the return of the growth rate (◇) to the growth rate in darkness is delayed. It is obvious from these data that the level of P_{fr} determines the growth rate in darkness over at least 6 hr following a 2-hr BL treatment.

The data in Figure 4 document another essential feature: when WL is given, no BL effect is detectable in weak WL. Only high light fluxes maintain a typical BL growth rate. At medium fluxes elongation rate returns gradually to the dark rate.

The simplest explanation of the data is that light absorbed by a separate BL photoreceptor is necessary to maintain responsivity to P_{fr}. With increasing age of the seedlings the re-

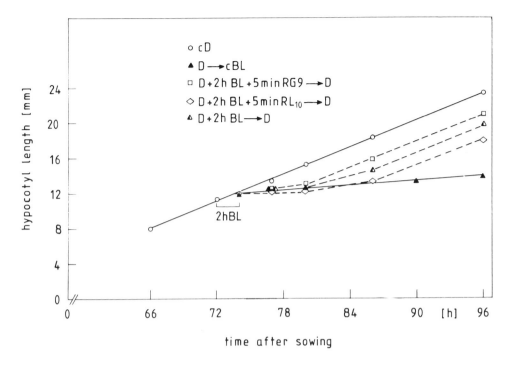

FIGURE 5. Time courses of hypocotyl elongation in the sesame seedling. Explication of symbols: cD, continuous darkness; cBL, continuous BL (7 W · m^{-2}, onset at 72 hr after sowing); → D, transfer to darkness; RG 9, long-wavelength far-red light, obtained with glass filter RG9 (10 W · m^{-2}, $\phi_{RG9} < 0.01$); RL$_{10}$, high-fluence-rate RL (6.8 W · m^{-2}, $\phi_{RL} = 0.8$). $\phi_{BL} = 0.38$. (Data of H. Drumm-Herrel and H. Mohr.)

quirement for BL increases strongly. On the other hand, brief light pulses — given to demonstrate photoreversibility of phytochrome — remain equally effective provided that responsivity to P$_{fr}$ exists.[21]

C. Synthesis of Plastidal GPD (Glyceraldehyde-3-Phosphate Dehydrogenase, EC 1.2.1.13) in the Shoot (Mainly Primary Leaf) of Milo (*Sorghum vulgare* Pers., cv. Weider Hybrid)

This case study about regulation of synthesis of a major enzyme of the chloroplast matrix documents the interdependent coaction of BL/UV and phytochrome in such photoresponses that are not related to flavonoid (anthocyanin) synthesis or growth.

It was found in the present case that responsivity towards P$_{fr}$ established by single light pulses is extremely weak in dark-grown milo shoots, while prolonged light treatments lead to a dramatic increase of responsivity (degree of response per unit P$_{fr}$).[22]

Figure 6 indicates that continuous light causes a rapid and strong responsivity increase ("responsivity amplification") which tends to saturate after approximately 6 hr. Blue and UV light are equally effective, and far more effective than RL. Since light-mediated changes of P$_{tot}$ levels are the same in RL, BL, and UV, it is clear that UV and BL cause a several times higher responsivity than RL.

VI. CONCLUSIONS DERIVED FROM THESE AND OTHER[16,17] CASE STUDIES

It is concluded that responsivity of a plant toward P$_{fr}$ strongly depends on the quality and quantity of the ambient light. A higher plant measures light throughout the spectrum and this information — obtained via phytochrome, cryptochrome, and UV-B photoreceptor — determines the efficiency of P$_{fr}$ action or, in other words, the actual responsivity towards P$_{fr}$ during growth and development.

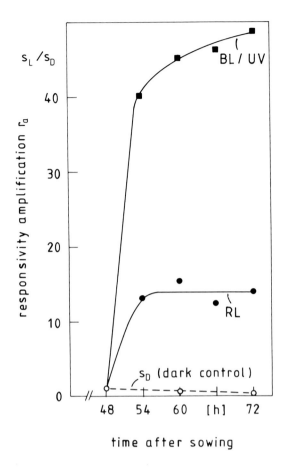

FIGURE 6.   Time course of responsivity amplification in continuous RL and BL/UV for phytochrome-mediated chloroplast glyceraldehyde-3-phosphate dehydrogenase (NADP dependent) induction in the milo shoot. The light treatment commenced 48 hr after sowing.[22] s_L, responsivity to P_{fr} in light-treated material; s_D, responsivity to P_{fr} in dark-grown material; S_D at 60 and 72 hr refers to s_D at 48 hr = 1.

As far as we know at present, the data in Figure 6 represent the usual interdependence of BL/UV and light absorbed by phytochrome: the short wavelength light strongly increases the effectivness of P_{fr}. However, in most cases studied so far, a BL/UV treatment is not *obligatory* for a P_{fr} action to occur. Rather, BL/UV causes an intensification of P_{fr}-mediated processes which occur even in RL alone, albeit at a low rate. It seems that the action of BL/UV and light absorbed by phytochrome during plant development must in general be conceived as an interdependent coaction as outlined in Table 2. It was shown recently[23] that the specific effect of BL/UV cannot be explained by an effect of light on gross protein synthesis. Rather, the pertinent data indicate that amplification of responsivity to P_{fr} by BL/UV is a specific process directly related to the mechanism of modulation of gene expression by phytochrome.

VII. THE PHENOMENON OF "BL-BLINDNESS" IN WEAK LIGHT

As pointed out in connection with Figure 4, no specific BL effect on hypocotyl growth is detectable in weak WL. Longitudinal growth of the axis in weak WL is precisely the

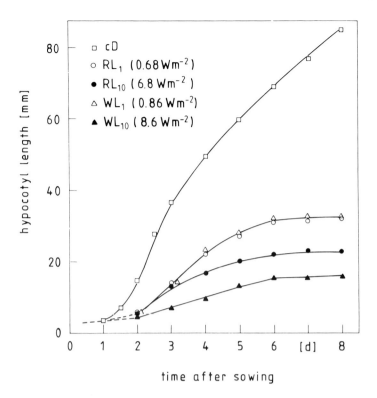

FIGURE 7. Time courses of hypocotyl elongation in the mustard seedling in continuous (c) RL and WL of two different fluence rates; cD, continuous darkness (Data of H. Drumm-Herrel.)

same as in RL of a corresponding fluence rate. This lack of a specific BL effect is amazing insofar as in dicotyledonous seedlings usually a very low fluence rate (e.g., 1 mW · m^{-2})[24] of continuous BL suffices to induce a strong phototropic growth response if applied *unilaterally*.[25] The apparent paradox was studied further using the mustard (*Sinapis alba* L.) seedling because hypocotyl longitudinal growth and phototropic bending in this seedling are known to be very sensitive towards light.[25] Moreover, it was shown previously that the phototropic response can under no circumstances be elicited by unilateral RL[26] while the seedling responds extremely sensitively towards unilateral BL. Figure 7 shows that hypocotyl longitudinal growth in the mustard seedling responds very strongly and in precisely the same way to low-fluence-rate RL and WL (an effect which can be attributed fully to light absorption by phytochrome). Only with increasing fluence rate, a transient but strong effect of BL comes into play which cannot be explained solely by the operation of phytochrome.

Regarding phototropism it was confirmed that the mustard seedlings respond to less than 1 mW · m^{-2} of continuous, unilateral BL or WL with a strong curvature.

From these kinds of data it is concluded that the BL-dependent *phototropic* growth response of a hypocotyl and the effect of BL on *longitudinal* growth of the hypocotyl are completely different phenomena.

ACKNOWLEDGMENTS

The research was supported by the Deutsche Forschungsgemeinschaft (SFB 206). I am greatly indebted to Dr. H. Drumm-Herrel, R. Oelmüller, and R. Schmidt for placing unpublished data at my disposal.

REFERENCES

1. **Mohr, H.,** Principles in plant morphogenesis, in *Axioms and Principles of Plant Construction,* Sattler, R., Ed., Martinus Nijhoff, The Hague, 93, 1982.
2. **Mohr, H.,** Criteria for photoreceptor involvement, in *Techniques in Photomorphogenesis,* Smith, H. and Holmes, M. G., Eds., Academic Press, London, 1984, chap. 2.
3. **Yatsuhashi, H., Hashimoto, T., and Shimizu, S.,** Ultraviolet action spectrum for anthocyanin formation in broom Sorghum first internode, *Plant Physiol.,* 70, 735, 1982.
4. **Mohr, H. and Drumm-Herrel, H.,** Coaction between phytochrome and blue/UV light in anthocyanin synthesis in seedlings, *Physiol. Plant.,* 58, 408, 1983.
5. **Drumm-Herrel, H. and Mohr, H.,** A novel effect of UV-B in a higher plant *(Sorghum vulgare), Photochem. Photobiol.,* 33, 391, 1981.
6. **Borthwick, H. A., Hendricks, S. B., Parker, M. W., Toole, E. H., and Toole, V. K.,** A reversible photoreaction controlling seed germination, *Proc. Natl. Acad. Sci. U.S.A.,* 38, 662, 1952.
7. **Rüdiger, W. and Scheer, H.,** Chromophores in photomorphogenesis, in *Encyclopedia of Plant Physiology, N.S.,* Vol. 16A, Shropshire, W. and Mohr, H., Eds., Springer-Verlag, Heidelberg, 1983, chap. 7.
8. **Pratt, L. H.,** Phytochrome: the protein moiety, *Annu. Rev. Plant Physiol.,* 33, 557, 1982.
9. **Schmidt, R. and Mohr, H.,** Evidence that a mustard seedling responds to the amount of P_{fr} and not to the P_{fr}/P_{tot} ratio, *Plant Cell Environ.,* 5, 495, 1982.
10. **Apel, K.,** Phytochrome-induced appearance of mRNA activity for the apoprotein of the light-harvesting chlorophyll a/b protein of barley *(Hordeum vulgare), Eur. J. Biochem.,* 97, 183, 1979.
11. **Tobin, E. M.,** Phytochrome-mediated regulation of messenger RNAs for the small subunit of ribulose-1.5-bisphosphate carboxylase and the light-harvesting chlorophyll a/b protein in *Lemna gibba, Plant Mol. Biol.,* 1, 35, 1981.
12. **Silverthorne, J. and Tobin, E. M.,** Demonstration of transcriptional regulation of specific genes by phytochrome action, *Proc. Nat. Acad. Sci. U.S.A.,* 81, 1112, 1984.
13. **Mösinger, E., Batschauer, A., Schäfer, E., and Apel, K.,** Phytochrome-control of in vitro transcription of specific genes in isolated nuclei from barley *(Hordeum vulgare), Eur. J. Biochem.,* 147, 137, 1985.
14. **Schäfer, E., Lassig, T. U., and Schopfer, P.,** Photocontrol of phytochrome destruction in grass seedlings. The influence of wave length and irradiance, *Photochem. Photobiol.,* 22, 193, 1975.
15. **Mohr, H., Drumm, H., Schmidt, R., and Steinitz, B.,** The effect of light pretreatments on phytochrome-mediated induction of anthocyanin and of phenylalanine ammonia lyase, *Planta,* 146, 369, 1979.
16. **Mohr, H.,** Interaction between blue light and phytochrome in photomorphogenesis, in *The Blue Light Syndrome,* Senger, H., Ed., Springer-Verlag, Berlin, 97, 1980.
17. **Mohr, H., Drumm-Herrel, H., and Oelmüller, R.,** Coaction of phytochrome and blue/UV light photoreceptors, in *Blue Light Effects in Biological Systems,* Senger, H., Ed., Springer-Verlag, Berlin, 6, 1984.
18. **Downs, R. J. and Siegelman, H. W.,** Photocontrol of anthocyanin synthesis in milo seedlings, *Plant Physiol.,* 38, 25, 1963.
19. **Mohr, H. and Drumm-Herrel, H.,** Interaction between blue/UV light and light operating through phytochrome in higher plants, in *Plants and the Daylight Spectrum,* Smith, H., Ed., Academic Press, London, 1981, 423.
20. **Drumm, H. and Mohr, H.,** The mode of interaction between blue (UV) light photoreceptor and phytochrome in anthocyanin formation of the Sorghum seedling, *Photochem. Photobiol.,* 27, 241, 1978.
21. **Drumm-Herrel, H. and Mohr, H.,** Mode of coaction of phytochrome and blue light photoreceptor in control of hypocotyl elongation, *Photochem. Photobiol.,* 40, 261, 1984.
22. **Oelmüller, R. and Mohr, H.,** Responsivity amplification by light in phytochrome-mediated induction of chloroplast glyceraldehyde-3-phosphate dehydrogenase (NADP-dependent, EC 1.2.1.13) in the shoot of milo *(Sorghum vulgare* Pers.), *Plant Cell Environ.,* 7, 29, 1984.
23. **Oelmüller, R. and Mohr, H.,** Specific action of blue light on phytochrome-mediated enzyme syntheses in the shoot of milo *(Sorghum vulgare* Pers.), *Plant Cell Environ.,* 8, 27, 1985.
24. **Steinitz, B. and Poff, K. L.,** Phototropism in Arabidopsis seedlings, *Plant Physiol.,* 75 (Suppl. no. 1), 73, 1984.
25. **Mohr, H.,** *Lectures on Photomorphogenesis,* Springer-Verlag, Heidelberg, 1972, chap. 22.
26. **Shropshire, W. and Mohr, H.,** Gradient formation of anthocyanin in seedlings of *Fagopyrum* and *Sinapis* unilaterally exposed to red and far-red light, *Photochem. Photobiol.,* 12, 145, 1970.
27. **Oelmüller, R. and Mohr, H.,** Mode of coaction between blue/UV light and light absorbed by phytochrome in light-mediated anthocyanin formation in the milo *(Sorghum vulgare* Pers.) seedling, *Proc. Natl. Acad. Sci. U.S.A.,* 82, 6124, 1985.

Chapter 11

THE RELATION OF PHOTOSYNTHESIS TO BLUE LIGHT EFFECTS

Donat-P. Häder

TABLE OF CONTENTS

I. INTRODUCTION

During evolution microorganisms have developed a number of very different receptor systems to respond to light. The blue light receptor is only one of them and is used especially by some flagellates for photomovement responses. In addition, heme proteins, stentorin, rhodopsins, and others are also utilized, to name but a few. Many prokaryotic and eukaryotic photosynthetic microorganisms make use of the pigments organized in the photosynthetic apparatus or a subset of these. The connection between photosynthesis and photomovement can be either on the energetic level (i.e., the light-dependent process depends on photophosphorylation) or it is linked to the electron transport chain. In addition to these clear-cut cases, a number of examples have been studied where only some of the accessory pigments are involved, but not chlorophyll. Also, other examples have been studied where the action spectra indicate the participation of the photosynthetic pigments but the photosynthetic electron transport chain does not seem to be involved.

II. PHOTOKINESIS

A. Prokaryotes

Many of the filamentous cyanobacteria glide when in contact with a substratum.[1-5] The mechanism of movement is not yet completely understood but several hypotheses have been suggested.[6] Since no cilia or flagella have been found, motility in gliding prokaryotes could be based on slime extrusion, undulating membranes, surface tension, or contracting fibrils.[7-9] Cytochalasin B inhibits gliding motility in mycoplasmas but not in cyanobacteria.[10]

The filaments of *Phormidium uncinatum* produce slime sheaths (Figure 1) in which the organisms move back and forth. This behavior has the advantage that the force can be generated all around the circumference rather than only along the small contact zone with the substratum. In addition, the slime sheath sticks to a variety of substrata, while the organism itself always moves in the same environment. Motility is based on a proton gradient across the cytoplasmic membrane.[11] This proton motive force can be generated by the photosynthetic or the respiratory pathway.[12] Some strains stop moving in darkness within a few seconds or minutes. Motility can be restored by irradiating only a few cells of a long filament with a light spot, indicating that there is an effective power transmission between cells.[13]

Cyanobacteria use light not only for energetic purposes but also as a source of information for finding a microenvironment with suitable light intensities.[14-16] The light-dependent responses are photokinesis, phototaxis, and photophobic responses (see Chapter 9, this volume). Photokinesis is the dependence of movement on the steady-state light intensity. After a change in light intensity, the organisms adapt their speed of movement to the new level within a few minutes (Figure 2).[17] There is a linear relationship between the velocity of a trichome and the logarithm of the illuminance up to an optimum value, beyond which the motility decreases again. The action spectra of cyanobacteria indicate the participation of the photosynthetic pigments. While some species utilize only chlorophyll as a photoreceptor pigment, others utilize in addition the accessory pigments, carotenoids, C-phycoerythrin, and C-phycocyanin (Figure 3).[12,17,18]

Inhibitor studies allow the separation of two types of responses. The inhibitor of the linear photosynthetic electron transport chain 3-(3,4-dichlorophenyl)-1,1 dimethyl urea (DCMU) impairs photokinesis in *Anabaena* drastically while *Phormidium, Lyngbia,* and *Nostoc* are either not inhibited at all or only at very high DCMU concentrations. The results indicate that in different cyanobacteria photokinesis is linked to either the cyclic photosynthetic electron transport involving only photosystem I (PS-I) or to both cyclic and noncyclic electron transport.[17,18] Uncouplers like carbonyl cyanide-*m*-chlorphenyl-hydrazon (CCCP), desaspidin, or dinitrophenol (DNP) drastically block photokinesis in all tested species. These results

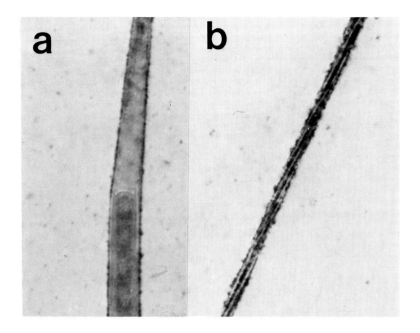

FIGURE 1. *Phormidium uncinatum* producing a slime sheath during forward locomotion (a) which collapses and is twisted when the filament leaves it (b) stained by china ink particles.

demonstrate that photokinesis is linked to photophosphorylation as a typical photocoupling process. Thus, the increased speed of movement is caused by an increased ATP production by the photosynthetic machinery.

Purple bacteria also show a photokinetic effect which had already been observed in 1883, by Engelmann.[19] The photokinetic action spectrum of *Rhodospirillum rubrum* shows a pronounced resemblance to the absorption spectrum, with a maximum near 510 nm and another at about 890 nm.[20] The maxima coincide with the absorption peaks of spirilloxanthin and bacteriochlorophyll. This result suggests that also in purple bacteria photokinesis is brought about by increased photosynthetic phosphorylation. This is supported by the fact that externally applied ATP increased the speed of movement up to sixfold.[20]

B. Eukaryotes

Photokinetic effects have been reported for a number of eukaryotic organisms. Bolte[21] observed that numerous flagellates, including *Euglena* and *Chlamydomonas,* stop swimming in high light intensities. Recently, *Euglena* has been reported to enhance its swimming speed in red light,[22] while Wolken and Shin[23] found an optimum in the blue spectral region around 460 nm and a smaller peak between 600 and 650 nm. This can indicate the participation of the chlorophylls, but the poor resolution of the action spectra do not allow a clear decision.

Diatoms also have an activity maximum in the red region of the spectrum absorbed by chlorophyll a, but the blue region is less effective than the absorption would suggest. In addition, also light absorbed by fucoxanthin between 500 and 550 nm induces a positive photokinetic effect.[24] Thus, also in diatoms, light could enhance speed by an increased photophosphorylation. The action spectrum of the motile unicellular rhodophyte *Porphyridium cruentum* resembles that of the cyanobacterium *Phormidium,* and the noncyclic photosynthetic electron transport seems to be responsible for the photokinetic effect.[17]

In desmids, photokinesis has been studied in more detail.[25] The action spectrum resembles the absorption spectrum (Figure 4).[26] The inhibitors of the photosynthetic electron transport chain, DCMU and 2,5-dibromo-3-methyl-6-isopropyl-*p*-benzoquinone (DBMIB), strongly impair photokinesis while they affect phototaxis and phobic responses to a much lesser

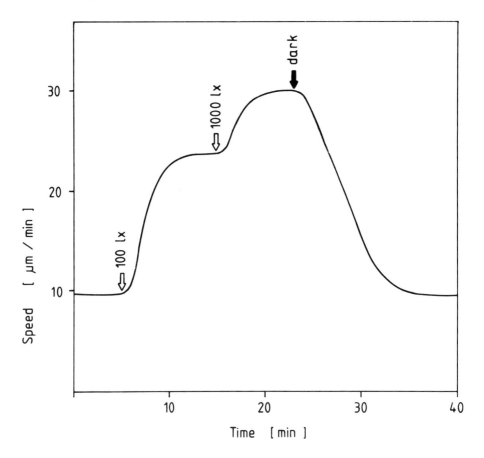

FIGURE 2. Photokinesis of *Anabaena variabilis*. Each change in light intensity (arrows) causes a gradual change in speed of movement which approaches a new steady state.[17]

degree (see below).[27] The uncouplers CCCP and DNP also strongly depress photokinesis.[28]

In summary, in all cases studied in any detail, the increase in the speed of movement by light seems to be brought about by a higher ATP synthesis in the photosynthetic pathway. This may even apply to ciliates with endosymbiontic green algae since photokinesis is impaired by DCMU.[29-32]

III. PHOTOTAXIS

Phototactic orientation, a directed movement with respect to the light direction, is found both in prokaryotes and eukaryotes.[33-38] While all prokaryotes use the photosynthetic pigments, or a subset thereof, only gliding eukaryotes seem to utilize the photosynthetic pigments for this purpose, while flagellated organisms have developed a specialized photoreceptor system (see Chapter 9).[39,40]

A. Cyanobacteria

Since purple bacteria, halobacteria, and other bacteria do not show phototaxis but only photophobic responses (which unfortunately have been mistermed "phototaxis" by some authors), we are restricted to cyanobacteria, which are the only prokaryotes capable of real phototactic orientation.

Most cyanobacteria show both positive and negative phototaxis while *Beggiatoa* and *Oscillatoria jenensis* have been reported to move exclusively away from the light source.[1]

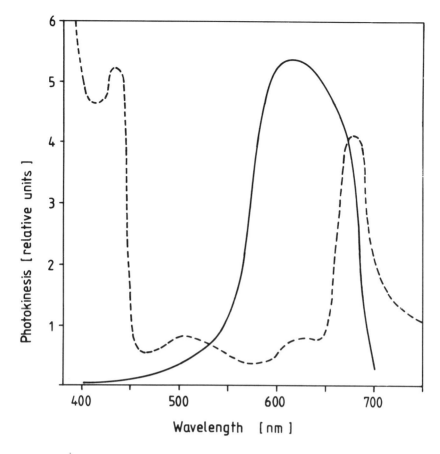

FIGURE 3. Photokinetic action spectra of *Phormidium ambiguum* (broken line) and *Anabaena variabilis* (solid line).[17]

Different cyanobacteria utilize two different strategies to move with respect to the light direction. The more primitive mechanism is used by *Oscillatoria* and *Phormidium:* the filaments reverse the direction of movement once every few minutes or hours. Upon unilateral illumination, those organisms which happen to move more or less parallel to the light rays prolong their path toward the light source and shorten the movement away from the light source (Figure 5a). Organisms perpendicular to the light beam are not affected but may change direction when encountering a mechanical obstacle so that eventually the whole population approaches the light source.[41-43] A true steering mechanism is used by the Nostocaceae. In contrast to *Phormidium*, the filaments of *Anabaena* do not rotate round their long axis during forward movement.[44] In lateral light the filaments turn actively into the direction of the incident light beam in low light intensities and away from the light source in higher intensities.[45] When a light field, incident from above, partially irradiates one half of a tip, it turns into the light field at low intensities and away at high intensities. Obviously the cells measure an internal light gradient and compare readings from the two sides to determine the light direction. It is interesting to note that each cell in a trichome can respond to light: when the light field is focused onto one half of some central cells the filament moves in a U-shaped figure with respect to the light direction.[44] This behavior is also found in lateral light (Figure 5b).

The phototactic action spectrum in *Phormidium* is different from that of photokinesis and the photophobic response: only the photosynthetic accesory pigments C-phycoerythrin, C-phycocyanin, and possibly carotenoids are effective, but chlorophyll a does not seem to be involved. In *Anabaena,* positive and negative phototaxis show a different spectral sensitivity

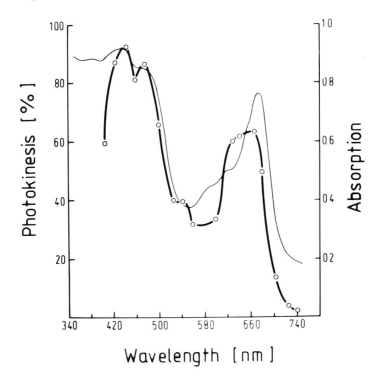

FIGURE 4.   Photokinetic action spectrum of the desmid *Cosmarium cucumis*. (From Wenderoth, K. and Häder, D.-P., *Planta,* 145, 1, 1979. With permission.)

(Figure 6), indicating that there are two different photoreceptor systems involved.[45] Positive phototaxis seems to be governed by the phycobilins and, to a smaller extent, by chlorophyll a. Negative phototaxis involves additional, yet unidentified, pigments absorbing around 500 nm and between 700 and 750 nm. Microscopical observation indicated that *Anabaena* orients itself perfectly well even in the presence of DCMU or DBMIB, which excludes a participation of the photosynthetic electron transport chain in the phototactic orientation.[45]

Negative phototaxis is sensitive to sodium azide. Application of NaN_3 reverses negative phototaxis into a positive one even at high fluence rates.[46] Sodium azide probably does not affect the photoperception but rather the reaction sign reversal generator. It is assumed that singlet oxygen, an active oxygen species developed under high fluence rates, controls the sign reversal. Singlet oxygen is quenched by azide. This hypothesis is further supported by the fact that other $^1O_2$ quenchers like L-histidin, 1,4-diazabicyclo[2.2.2]-octane (DABCO) and potassium iodide, as well as gasing with nitrogen, also inhibit negative phototaxis.[45,47,48]

B. Eukaryotes

Gliding desmids show phototactic orientation in extremely low light intensities on the order of 10^{-4} to 10^{-5} lx. The thresholds follow a circadian rhythm and are about 100 times higher during the night.[49,50] The action spectra of *Micrasterias* and other species strongly resemble the absorption spectrum with maxima between 370 and 470 nm and a second peak around 660 nm.[25-27] Thus, the action spectra indicate the activity of the photosynthetic apparatus. However, neither inhibitors of the photosynthetic electron transport chain nor uncouplers affect phototactic orientation. As long as the algae are able to move — higher concentrations of these drugs impair motility — they are perfectly oriented with respect to the light direction.[27,28] Thus, we are faced with the puzzling situation that chlorophyll seems to be the primary photoreceptor of the response but that the photosynthetic electron transport chain does not seem to be involved.

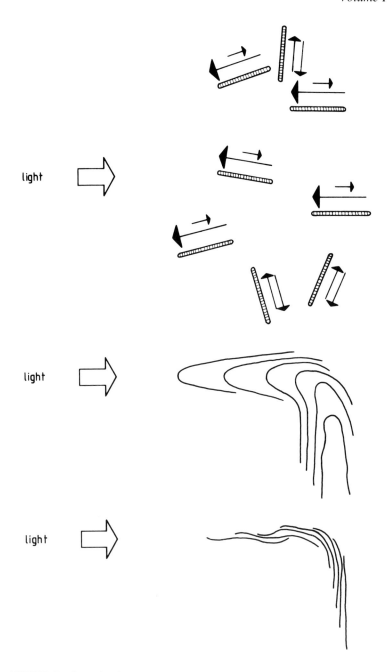

FIGURE 5. Strategies for phototactic orientation of cyanobacteria in lateral light (arrows). For details see text.[16,45]

The motile unicellular red alga *Porphyridium* has a phototactic threshold of 1 lx, and at 100 lx the response is already saturated. The action spectrum indicates a maximum at 443 nm and a second one above 500 nm ($>10^{-11}$ mol cm^{-2} sec^{-1}). In low photon fluence rates (10^{-12} to 10^{-13} mol cm^{-2} sec^{-1}) the phototactic orientation decreases, and the spectral peaks move closer together and eventually form one maximum in the blue region of the spectrum (443 nm).[51] The photoreceptor pigment has neither been identified in this organism nor in the also gliding flagellate *Euglena mutabilis*.[52]

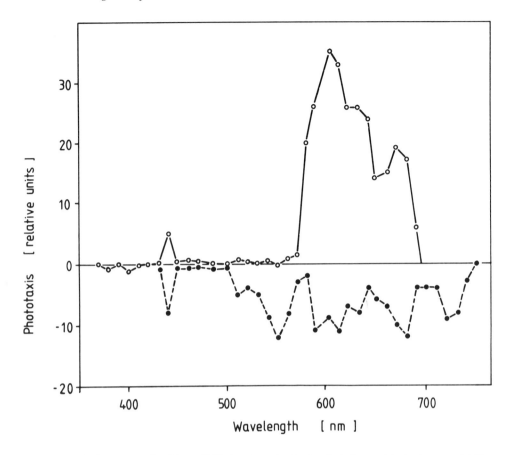

FIGURE 6.   Action spectra of positive (solid line) and negative (broken line) phototaxis in *Anabaena variabilis*. (From Nultsch, W., Schuchart, H., and Höhl, M., *Arch. Microbiol.*, 122, 85, 1979. With permission.)

IV. PHOTOPHOBIC RESPONSES

A sudden change in light intensity (step-up or step-down) causes a transient photophobic response. Both temporal and spatial changes can induce a specific reaction, the form of which depends on the organism under observation.

A. Purple Bacteria

Engelmann[19] observed purple bacteria which reversed the direction of movement each time they entered a dark field. This reversal was abrupt, as if the cells were "afraid" of the dark area (*phobos* = Greek for fright). The action spectrum coincides with that of photokinesis and the absorption spectrum of this organism. During the phobic response the membrane potential of the cell changes. This has recently been found by following the absorption bandshift of native carotenoids or artificially applied oxanol dyes which are indicators for the electrical potential across the membrane.[53,54]

The internal potential rises during light (or chemical) stimulation. This effect can be cancelled by 10^{-7} *M* CCCP. Mutants which lack the photosynthetic reaction center do not show light-induced phobic responses, but chemotaxis is not impaired. Thus, obviously the photosynthetic electron transport chain is responsible for photoperception in purple bacteria. The same is true for the photophobic response in cyanobacteria.

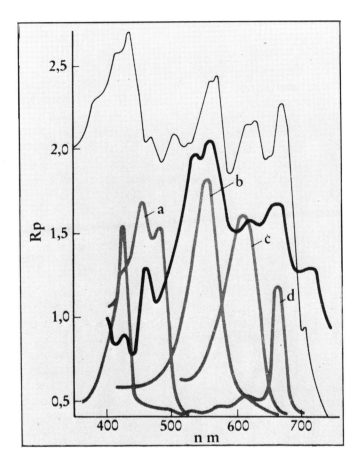

FIGURE 7. Action spectrum of photophobic responses in *Phormidium uncinatum* (thick line) in comparison with the in vivo low temperature ($-196°C$) absorption spectrum (fine line) and the absorption spectra of β-carotene (a), C-phycoerythrin (b), C-phycocyanin (c) and chlorophyll a (d). (From Häder, D.-P., *BIUZ*, 14, 78, 1984. With permission.)

B. Cyanobacteria

The filamentous cyanobacterium, *Phormidium uncinatum* stops after a delay of a few seconds upon a sudden decrease in light intensity. After a resting period also of a few seconds it glides back on its original track. This response can be easily demonstrated by projecting light fields into a homogeneous suspension of filaments. During their random movement in darkness some organisms enter the light field. The transition from the dark surrounding field into the irradiated zone does not cause a phobic response. However, when the organisms try to leave the light field again, they reverse their direction each time they enter the dark field. Thus, the filaments accumulate over a period of several hours. This technique can be used to determine the threshold values and the spectral sensitivity of the organisms. The zero threshold was found at 0.03 lx, which is well below moonlight intensity.[55] The organisms respond not only at a dark/light boundary, but also upon moving from a brighter to a weaker light field; the difference threshold is about 4%.[56]

1. Photoperception

The action spectrum of the photophobic response measured using the light-trap technique described above extends throughout the visible spectrum and shows a remarkable resemblance to the low-temperature absorption spectrum (Figure 7). The comparison with the extracted pigments indicates that chlorophyll a, carotenoids, and the phycobilins C-phycoerythrin and C-phycocyanin are involved in the photoperception,[57] which suggests that photophobic response in *Phormidium* is coupled to photosynthesis.

In contrast to photokinesis, photophobic responses in cyanobacteria are not coupled to photophosphorylation since uncouplers do not inhibit the response. Inhibitors of the photosynthetic electron transport chain, however, drastically impair photophobic reactions. In *Phormidium,* the two photosystems can be activated separately using appropriate spectral bands. DCMU blocks specifically phobic responses when photosystem II (PS II) is activated while PS I is unaffected. DBMIB, on the other hand, impairs PS I reactions while PS II responses are not inhibited (Figure 8).[58-62] These results indicate that plastoquinone plays a key role in photoperception, since it is the only known redox component between the DCMU and DBMIB inhibition sites.[63]

Plastoquinone has a central position in the photosynthetic electron transport chain since it conducts a vectorial proton transport across the cytoplasmic membrane. During the electron flux along the redox system transport chain, protons are taken up at the outer side of the thylakoids and are subsequently released into the interior of the thylakoid vesicles. In cyanobacteria the thylakoids are closed sacks which are not enclosed by an additional plastid membrane. Therefore, the protons are transported from the cytoplasmic compartment into the thylakoids, which produces a gradient of up to 2 to 3 pH units.[64] When a trichome moves into a shaded area this proton gradient will break down in the front cells which could be sensed and serve as a signal for the photophobic transduction chain.

This hypothesis is supported by the fact that protonophores, which break down the proton gradient, inhibit the phobic response. Furthermore, reversals can be induced by a suitable pH jump in darkness. The filaments usually buffer their medium to pH 7.2. When an external buffer with a pH of ≤4.8 ≥12.4 is added, the filaments stop moving. Most organisms continue to move in the same direction after addition of a buffer with a pH between 5.6 and 12.1. In the range from 4.9 to 5.6, however, most filaments reverse their movement.[65] This result can be interpreted by assuming that the pH change stimulated the phobic transduction chain bypassing the photoreceptor.

The generation of the proton gradient in light and the subsequent breakdown in darkness also causes electrical potential changes in the cytoplasmic compartment since positive charges are transported across the membrane. In fact, electrical potential changes of a few millivolts can be measured with internal microglass electrodes when the cells are subjected to a light/dark alternation.[66] The action spectrum of the light-induced potential changes resembles that of photophobic responses. When these potential changes are inhibited by applying the lipophilic cation triphenylmethylphosphonium$^-$ (TPMP$^-$Br$^+$), phobic responses are also impaired. These ions can easily penetrate the membrane, follow passively an existing electrical gradient, and thus decrease the membrane potential.

2. Signal Amplification

Since the organisms respond to very low light intensities and also to small changes in intensity, these signals may not produce very large electrical potential changes. Therefore, a considerable signal amplification has to take place in the sensory transduction chain. This is often brought about by gated cation currents.[68-73] Channels are protein macromolecules located in the membrane; they have two distinct features — they are specific for certain ions and some can be opened and closed by electrical potentials across the membrane. The existence of voltage-dependent, cation-specific channels and their role in the sensory transduction chain can be demonstrated by several techniques. First of all, the channels can be blocked by specific blockers. Lanthanum ions and ruthenium red inhibit specifically calcium channels and impair the photophobic response.[76] Furthermore, the naturally existing channels can be bypassed by artificial channels called ionophores. Application of valinomycin or

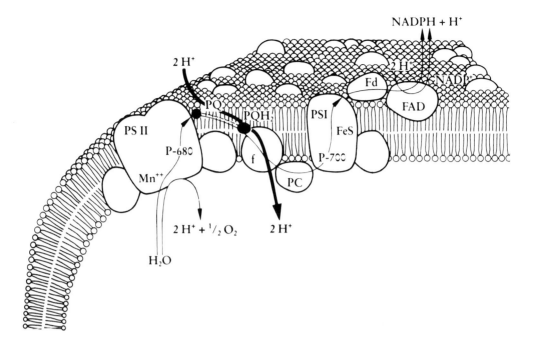

FIGURE 8.   Sector from the thylakoid membrane with the components of the photosynthetic electron transport chain. Plastoquinone takes up protons from the cytoplasmic compartment and releases them to the inside of the vesicle. (From Häder, D.-P., *BIUZ*, 14, 78, 1984. With permission.)

gramicidin, ionophores for monovalent cations, had no effect on phobic responses in *Phormidium,* but calcimycin (A23187), a specific ionophore for calcium, blocked the response drastically.[76-78] Removal of Ca^{2+} from the medium also abolished the response and subsequent addition of external calcium restored it.[79] Thus, the sensory transduction chain seems to involve gated calcium currents, which have also been found to be involved in sensory responses of a number of different organisms.[80-83]

The direction of the calcium fluxes can be revealed by the use of radioactively labeled $^{45}Ca^{2+}$. Shortly after a decrease in light intensity the cells take up labeled calcium from the outside and extrude it during the following seconds.[79] Thus, a step-down in light intensity causes a breakdown of a previously established proton gradient which results in a small decrease of the electrical resting potential. This event causes calcium channels to open which allows a massive calcium influx along a previously established calcium gradient (by active calcium pumps, Ca^{2+}-ATPase), which further decreases the membrane potential (Figure 9).[84-88]

The filaments of *Phormidium* do not show a morphological polarity; i.e., front and rear end are indistinguishable. The temporary direction of movement is established by an electrical gradient between front and rear end. The leading tip is more negative than the rear end. This electrical gradient is reversed during a photophobic response due to the depolarization of the former front end by the massive calcium influx.[89,90] The involvement of ionic currents in the transduction chains of light-triggered responses has been demonstrated also in other bacterial systems.[91-93]

Light responses control the movement of cyanobacteria in their natural environment. The step-down photophobic response prevents the organisms from entering dark areas.[94] Recently, a second phobic response has been found which is caused by a step-up in light intensity.[86,90] This reaction prevents the organisms from moving into fields of too bright light intensities which would photooxidize their pigments and eventually kill the cells.

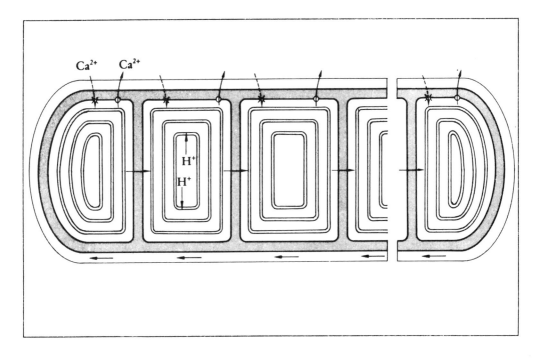

FIGURE 9.   Model to explain photophobic responses in *Phormidium:* small electrical potential changes, caused by the generation of the proton gradient over the thylakoid membrane in light and subsequent breakdown in darkness trigger specific cation channels which allow a massive Ca^{2+} influx. Afterwards Ca^{2+} is actively pumped out of the cell. (From Häder, D.-P., *BIUZ*, 14, 78, 1984. With permission.)

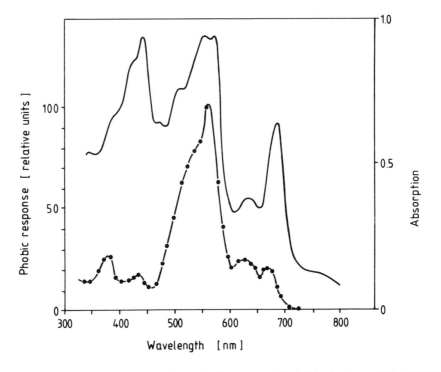

FIGURE 10.   Photophobic action spectrum of *Porphyridium cruentum* (circles) in a background light of 0.5 Wm^{-2} (BG 12) compared to the in vivo absorption spectrum (continuous line). (From Schuchart, H., *Arch. Microbiol.*, 128, 105, 1980. With permission.)

C. Eukaryotes

The unicellular red alga *Porphyridium cruentum* changes direction of movement and turns back when entering a dark field.[95] The action spectrum indicates the participation of both photosynthetic photosystems represented by the chlorophylls, phycobilins, and carotenoids (Figure 10). When the wavelength of a background irradiation is varied the action spectrum can be altered. Exciting photosystems I or II, respectively, causes an effective spillover of light energy from photosystem II to photosystem I.

Diatoms also show photophobic responses. Judging from the action spectra there are two different types: *Navicula* uses the chlorophylls a and c, carotenes, and fucoxanthin, the main accessory pigment of the organism. Thus, phobic responses in *Navicula* are controlled by the photosynthetic apparatus.[96,97] The related genus *Nitzschia*, however, shows a blue light action spectrum, while red light is not effective. The photoreceptor pigments for this type have not yet been identified.

REFERENCES

1. **Häder, D.-P.,** Photomovement, in *Encyclopedia of Plant Physiology,* N.S. Vol. 7, *Physiology of Movements,* Haupt, W. and Feinleib, M. E., Eds., Springer-Verlag, Berlin, 1979, 268.
2. **Raven, J. A.,** Cyanobacterial motility as a test of the quantitative significance of proticity transmission along membranes, *New Phytol.,* 94, 511, 1983.
3. **Nultsch, W.,** Photomotile responses in gliding organisms and bacteria, in *Photoreception and Sensory Transduction in Aneural Organisms,* Lenci, F. and Colombetti, G., Eds., Plenum Press, New York, 1980, 69.
4. **Häder, D.-P.,** Wie orientieren sich Cyanobakterien im Licht?, *BIUZ,* 14, 78, 1984.
5. **Poff, K. L. and Hong, C. B.,** Photomovement and photosensory transduction in microorganisms, *Photochem. Photobiol.,* 36, 749, 1982.
6. **Halfen, L. N.,** Gliding movements, in *Encyclopedia of Plant Physiology,* N.S. Vol. 7, *Physiology of Movements,* Haupt, W. and Feinleib, M. E., Eds., Springer-Verlag, Berlin, 1979, 250.
7. **Dickson, M. R., Kouprach, S., Humphrey, B. A., and Marshall, K. C.,** Does gliding motility depend on undulating membranes?, *Micron,* 11, 381, 1980.
8. **Keller, K. H., Grady, M., and Dworkin, M.,** Surface tension gradients: feasible model for gliding motility of *Myxococcus xanthus, J. Bacteriol.,* 155, 1358, 1983.
9. **Dworkin, M., Keller, K. H., and Weisberg, D.,** Experimental observations consistent with a surface tension model of gliding motility of *Myxococcus xanthus, J. Bacteriol.,* 155, 1367, 1983.
10. **Maniloff, J. and Chaudhuri, U.,** Gliding mycoplasmas are inhibited by cytochalasin B and contain a polymerizable protein fraction, *J. Supramol. Struct.,* 12, 299, 1979.
11. **Glagoleva, T. N., Glagolev, A. N., Gusev, M. V., and Nikitina, K. A.,** Protonmotive force supports gliding in cyanobacteria, *FEBS Lett.,* 117, 49, 1980.
12. **Nultsch, W. and Häder, D.-P.,** Light perception and sensory transduction in photosynthetic prokaryotes, *Struct. Bonding (Berlin),* 41, 111, 1980.
13. **Chailakhyan, L. M., Glagolev, A. N., Glagoleva, T. N., Murvanidze, G. V., Potapova, T. V., and Skulachev, V. P.,** Intercellular power transmission along trichomes of cyanobacteria, *Biochim. Biophys. Acta,* 679, 60, 1982.
14. **Fay, P.,** *The Blue-Greens,* The Institute of Biology's Studies in Biology No. 160, Edward Arnold, London, 1983.
15. **Castenholz, R. W.,** Motility and taxes, in *The Biology of Cyanobacteria,* Botanical monographs, Vol. 19, Carr, N. G. and Whitton, B. A., Eds., Blackwell Scientific, Oxford, 1982, 413.
16. **Tevini, M. and Häder, D.-P.,** *Allgemeine Photobiologie,* Thieme, Stuttgart, 1985.

17. **Nultsch, W.,** Der Einfluß des Lichtes auf die Bewegung phototropher Mikroorganismen. I. Photokinesis, *Abh. Marburger Gelehrten Ges.,* 2, 143, 1974.
18. **Nultsch, W. and Hellmann, W.,** Untersuchungen zur Photokinesis von *Anabaena* var., *Arch. Microbiol.,* 82, 76, 1972.
19. **Engelmann, T. W.,** *Bakterium photometricum.* Ein Beitrag zur vergleichenden Physiologie des Licht- und Farbensinnes, *Pflüger's Arch.,* 30, 95, 1883.
20. **Throm, G.,** Untersuchungen zum Reaktionsmechanismus von Phobotaxis und Kinesis an *Rhodospirillum rubrum, Arch. Protistenkd.,* 110, 313, 1968.
21. **Bolte, E.,** Über die Wirkung von Licht und Kohlensäure auf die Beweglichkeit grüner und farbloser Schwärmzellen, *Jahrb. Wiss. Bot.,* 59, 287, 1920.
22. **Zhenan, M. and Shouyu, R.,** The effect of red light on photokinesis of *Euglena gracilis,* in *Proc. Joint China-U.S. Phycology Symp.,* Tseng, C. K., Ed., Science Press, Beijing, 1983, 311.
23. **Wolken, J. J. and Shin, E.,** Photomotion in *Euglena gracilis.* I. Photokinesis, *J. Protozool.,* 5, 39, 1958.
24. **Nultsch, W.,** Phototactic and photokinetic action spectra of the diatom *Nitzschia communis, Photochem. Photobiol.,* 14, 705, 1971.
25. **Häder, D.-P. and Wenderoth, K.,** Role of three basic light reactions in photomovement of desmids, *Planta,* 137, 207, 1977.
26. **Wenderoth, K. and Häder, D.-P.,** Wavelength dependence of photomovement in desmids, *Planta,* 145, 1, 1979.
27. **Häder, D.-P.,** Effects of inhibitors on photomovement in desmids, *Arch. Microbiol.,* 129, 168, 1981.
28. **Häder, D.-P.,** Coupling of photomovement and photosynthesis in desmids, *Cell Motil.,* 2, 73, 1982.
29. **Reisser, W.,** Host-symbiont interaction in *Paramecium bursaria:* physiological and morphological features and their evolutionary significance, *Ber. Dtsch. Bot. Ges.,* 94, 557, 1982.
30. **Reisser, W. and Häder, D.-P.,** Role of endosymbiotic algae in photokinesis and photophobic responses of ciliates, *Photochem. Photobiol.,* 39, 673, 1984.
31. **Pado, R.,** The effect of white light on kinesis in the protozoans *Paramecium bursaria, Acta Protozool.,* 14, 83, 1975.
32. **Niess, D., Reisser, W., and Wiessner, W.,** The role of endosymbiotic algae in photoaccumulation of green *Paramecium bursaria, Planta,* 152, 268, 1981.
33. **Häder, D-P.,** Photomovement, in *Blue Light Effects in Biological Systems,* Senger, H., Ed., Springer-Verlag, Berlin, 1984, 435.
34. **Halldal, P.,** Light and microbiol activities, in *Contemporary Microbial Ecology,* Ellwood, D. C., Hedger, J. N., Latham, M. J., and Lynch, J. M., Eds., Academic Press, New York, 1980, 1.
35. **Diehn, B.,** Experimental determination and measurement of photoresponses, in *Photoreception and Sensory Transduction in Aneural Organisms,* Lenci, F. and Colombetti, G., Eds., Plenum Press, New York, 1980, 107.
36. **Kivic, P. A. and Walne, P. L.,** Algal photosensory apparatus probably represent multiple parallel evolutions, *Biosystems,* 16, 31, 1983.
37. **Colombetti, G. and Lenci, F.,** Identification and spectroscopic characterization of photoreceptor pigments, in *Photoreception and Sensory Transduction in Aneural Organisms,* Lenci, F. and Colombetti, G., Eds., Plenum Press, New York, 1980, 173.
38. **Haupt, W.,** Orientierungsbewegungen bei niederen Pflanzen und Bakterien. Erkenntnisse der biologischen Forschung, *Universitas, Ztschr. Wiss. Kunst, Lit.,* 36, 531, 1981.
39. **Nultsch, W.,** The photocontrol of movement of *Chlamydomonas,* in *The Biology of Photoreception,* Cosens, D. J. and Vince-Price, D., Eds., Symp. Soc. Exp. Biol. No. 36, Cambridge University Press, Cambridge, 1983, 521.
40. **Nultsch, W., Throm, G., and Rimscha, J.,** Phototaktische Untersuchungen an *Chlamydomonas reinhardii* Dangeard in homokontinuierlicher Kultur, *Arch. Microbiol.,* 80, 351, 1971.
41. **Nultsch, W.,** Phototaxis and photokinesis, in *Primitive Sensory and Communication Systems,* Carlile, M. J., Ed., Academic Press, New York, 1975, 29.
42. **Burkart, U. and Häder, D.-P.,** Phototactic attraction in light trap experiments: a mathematical model, *J. Math. Biol.,* 10, 257, 1980.
43. **Häder, D.-P.,** Computer-based evaluation of phototactic orientation in microorganisms, *EDV. Med. Biol.,* 12, 27, 1981.
44. **Nultsch, W. and Wenderoth, K.,** Partial irradiation experiments with *Anabaena variabilis* (Kütz.), *Z. Pflanzenphysiol.,* 111, 1, 1983.
45. **Nultsch, W., Schuchart, H., and Höhl, M.,** Investigations on the phototactic orientation of *Anabaena variabilis, Arch. Microbiol.,* 122, 85, 1979.
46. **Nultsch, W., Schuchart, H., and Koenig, F.,** Effects of sodium azide on phototaxis of the blue-green alga *Anabaena variabilis* and consequences to the two-photoreceptor systems-hypothesis, *Arch. Microbiol.,* 134, 33, 1983.
47. **Matheson, I. B. C., Etheridge, R. D., Kratowich, N. R., and Lee, J.,** The quenching of singlet oxygen by amino acids and proteins, *Photochem. Photobiol.,* 21, 165, 1975.

48. **Schuchart, H. and Nultsch, W.,** Possible role of singlet molecular oxygen in the control of the phototactic reaction sign of *Anabaena variabilis, J. Photochem.,* 25, 317, 1984.

49. **Neuscheler, W.,** Bewegung und Orientierung bei *Micrasterias denticulata* Breb. im Licht. I. Zur Bewegungs- and Orientierungsweise, *Z. Pflanzenphysiol.,* 57, 46, 1967.

50. **Neuscheler, W.,** Bewegung und Orientierung bei *Micrasterias denticulata* Breb. im Licht. II. Photokinesis und Phototaxis, *Z. Pflanzenphysiol.,* 57, 151, 1967.

51. **Nultsch, W. and Schuchart, H.,** Photomovement of the red alga *Porphyridium cruentum* (Ag.) Naegeli. II. Phototaxis, *Arch. Microbiol.,* 125, 181, 1980.

52. **Häder, D.-P. and Melkonian, M.,** Phototaxis in the gliding flagellate, *Euglena mutabilis, Arch. Microbiol.,* 135, 25, 1983.

53. **Armitage, J. P. and Evans, C. W.,** Comparison of the carotenoid bandshift and oxanol dyes to measure membrane potential changes during chemotactic stimulation of *Rhodopseudomonas sphaeroides* and *Escherichia coli, FEBS Lett.,* 126, 98, 1981.

54. **Armitage, J. P. and Evans, M. C. W.,** The reaction centre in the phototactic and chemotactic response of photosynthetic bacteria, *FEMS Microbiol. Lett.,* 11, 89, 1981.

55. **Nultsch, W.,** Der Einfluß des Lichtes auf die Bewegung der Cyanophyceen. III. Photophobotaxis bei *Phormidium uncinatum, Planta,* 58, 647, 1962.

56. **Nultsch, W. and Häder, D.-P.,** Bestimmungen der photo-phobotaktischen Unterschiedsschwelle bei *Phormidium uncinatum, Ber. Dtsch. Bot. Ges.,* 83, 185, 1970.

57. **Glazer, A. N.,** Phycobilisome. A macromolecular complex optimized for light energy transfer, *Biochim. Biophys. Acta,* 768, 29, 1984.

58. **Häder, D.-P. and Nultsch, W.,** Negative photo-phobotactic reactions in *Phormidium uncinatum, Photochem. Photobiol.,* 18, 311, 1973.

59. **Häder, D.-P.,** Participation of two photosystems in the photophobotaxis of *Phormidium uncinatum, Arch. Microbiol.,* 96, 255, 1974.

60. **Häder, D.-P.,** The effect of inhibitors on the electron flow triggering photophobic reactions in Cyanophyceae, *Arch. Microbiol.,* 103, 169, 1975.

61. **Häder, D.-P.,** Further evidence for the electron pool hypothesis. The effect of KCN and DSPD on the photophobic reaction in the filamentous blue-green alga *Phormidium uncinatum, Arch. Microbiol.,* 110, 301, 1976.

62. **Häder, D.-P.,** Phobic reactions between two adjacent monochromatic light fields, *Z. Pflanzenphysiol.,* 78, 173, 1976.

63. **Häder, D.-P. and Poff, K. L.,** Spectrophotometric measurement of plastoquinone photoreduction in the blue-green alga, *Phormidium uncinatum, Arch. Microbiol.,* 131, 347, 1982.

64. **Padan, E. and Schuldiner, S.,** Energy transduction in the photosynthetic membranes of the cyanobacterium (blue-green alga) *Plectonema boryanum, J. Biol. Chem.,* 253, 3281, 1978.

65. **Häder, D.-P.,** Electrical and proton gradients in the sensory transduction of photophobic responses in the blue-green alga, *Phormidium uncinatum, Arch. Microbiol.,* 130, 83, 1981.

66. **Häder, D.-P.,** Extracellular and intracellular determination of light-induced potential changes during photophobic reactions in blue-green algae, *Arch. Microbiol.,* 119, 75, 1978.

67. **Häder, D.-P.,** Effect of inhibitors and uncouplers on light-induced potential changes triggering photophobic responses, *Arch. Microbiol.,* 120, 57, 1979.

68. **Goll, A., Ferry, D. R., and Glossmann, H.,** Target size analysis and molecular properties of Ca^{2+} channels labelled with [^{3}H]verapamil, *Eur. J. Biochem.,* 141, 177, 1984.

69. **Prusti, R. K., Song, P.-S., Häder, D.-P., and Häder, M.,** Caffeine-enhanced photomovement in the ciliate, *Stentor coeruleus, Photochem. Photobiol.,* 40, 369, 1984.

70. **Marme, D. and Dieter, P.,** Role of Ca^{2+} and calmodulin in plants, in *Calcium and Cell Function,* Vol. 4, Cheung, W. Y., Ed., Academic Press, New York, 1983, 264.

71. **Goldworthy, A.,** The evolution of plant action potentials, *J. Theor. Biol.,* 103, 645, 1983.

72. **Simons, P. J.,** The role of electricity in plant movements, *New Phytol.,* 87, 11, 1981.

73. **Kim, I.-H., Prusti, R. K., Song, P.-S., Häder, D.-P., and Häder, M.,** Phototaxis and photophobic responses in *Stentor coeruleus.* Action spectrum and role of Ca^{2+} fluxes, *Biochim. Biophys. Acta,* 799, 298, 1984.

74. **Häder, D.-P.,** Photosensory transduction chains in procaryotes, in *Photoreception and Sensory Transduction in Aneural Organisms,* Lenci, F. and Colombetti, G., Eds., Plenum Press, New York, 1980, 355.

75. **Hess, P. and Tsien, R. W.,** Mechanism of ion permeation through calcium channels, *Nature (London),* 309, 453, 1984.

76. **Häder, D.-P.,** Gated ion fluxes involved in photophobic responses of the blue-green alga, *Phormidium uncinatum, Arch. Microbiol.,* 131, 77, 1982.

77. **Murvanidze, G. V. and Glagolev, A. N.,** Calcium ions regulate reverse motion in phototactically active *Phormidium uncinatum,* and *Halobacterium halobium, FEMS Microbiol. Lett.,* 12, 3, 1981.

78. **Murvanidze, G. V.,** The role of Ca^{2+} and the membrane potential in the formation of the taxis signal in cyanobacteria, *Bull. Acad. Sci., Georgian (SSR),* 104, 173, 1981.

79. **Häder, D.-P. and Poff, K. L.,** Dependence of the photophobic response of the blue-green alga, *Phormidium uncinatum*, on cations, *Arch. Microbiol.*, 132, 345, 1982.

80. **Baryshev, V. A.,** Regulation of *Halobacterium halobium* motility by Mg^{2+} and Ca^{2+} ions, *FEMS Microbiol. Lett.*, 14, 139, 1982.

81. **Abeliovich, A. and Gan, J.,** Site of Ca^{2+} action in triggering motility in the cyanobacterium *Spirulina subsalsa, Cell Motil.*, 2, 393, 1982.

82. **Williamson, R. E. and Ashley, C. C.,** Free Ca^{2+} and cytoplasmic streaming in the alga *Chara, Nature (London)*, 296, 647, 1982.

83. **Murvanidze, G. V. and Glagolev, A. N.,** Role of Ca concentration gradient and the electrochemical potential of H^+ in the reaction of phototaxis of cyanobacteria, *Biophysics*, 28, 887, 1983.

84. **Murvanidze, G. V. and Glagolev, A. N.,** Electrical nature of the taxis signal in cyanobacteria, *J. Bacteriol.*, 150, 239, 1982.

85. **Häder, D.-P. and Burkart, U.,** Mathematical simulation of photophobic responses in blue-green algae, *Math. Biosci.*, 58, 1, 1982.

86. **Häder, D.-P. and Burkart, U.,** Enhanced model for photophobic responses of the blue-green alga, *Phormidium uncinatum, Plant Cell Physiol.*, 23, 1391, 1982.

87. **Murvanidze, G. V., Gabai, V. L., and Glagolev, A. N.,** The role of membrane potential in the photophobic response and certain aspects of chemotaxis in the cyanobacterium *Phormidium uncinatum, Academic NAUK CCCP Microbiologia*, 51, 240, 1982.

88. **Glagolev, A. N.,** *Motility and Taxis in Prokaryotes*, Physicochem. Biol. Series, Vol. 3, Harwood, Chur, Switzerland, 1983.

89. **Murvanidze, G. V., Gabai, V. L., and Glagolev, A. N.,** Taxic responses in *Phormidium uncinatum, J. Gen. Microbiol.*, 128, 1623, 1982.

90. **Häder, D.-P.,** Effects of UV-B on motility and photoorientation in the cyanobacterium, *Phormidium unicinatum, Arch. Microbiol.*, 140, 34, 1984.

91. **Brown, I. I., Galperin, M. Y., Glagolev, A. N., and Skulachev, V. P.,** Utilization of energy stored in the form of Na^+ and K^+ ion gradients by bacterial cells, *Eur. J. Biochem.*, 134, 345, 1983.

92. **Baryshev, V. A., Glagolev, A. N., and Skulachev, V. P.,** Sensing of H^+ in phototaxis of *Halobacterium halobium, Nature (London)*, 292, 338, 1981.

93. **Hildebrand, E.,** Comparative discussion of photoreception in lower and higher organisms. Structural and functional aspects, in *Photoreception and Sensory Transduction in Aneural Organisms*, Lenci, F. and Colombetti, G., Eds., Plenum Press, New York, 1980, 319.

94. **Nelson, D. C. and Castenholz, R. W.,** Light responses of *Beggiatoa, Arch. Microbiol.*, 131, 146, 1982.

95. **Schuchart, H.,** Photomovement of the red alga *Porphyridium cruentum* (Ag.) Naegeli. III. Action spectrum of the photophobic response, *Arch. Microbiol.*, 128, 105, 1980.

96. **Wenderoth, K.,** Photophobische Reaktionen von Diatomeen im monochromatischen Licht, *Ber. Dtsch. Bot. Ges.*, 92, 313, 1979.

97. **Wenderoth, K.,** Reaktion der Kieselalge *Navicula peregrina* auf Belichtung verschiedener Zellabschnitte, *Inst. Wiss. Film*, Biol. 16/21—C 1466.

Index

INDEX